# Green Bites

©John Vlahakis

All rights reserved. No part of this book may be reproduced in any form by an electronic or mechanical means (including photocopying, recording, or information storage and retrieval) without permission in writing from the publisher.

Earthy, LLC books may be purchased at special quantity discounts for business or sales promotional use. For information, please e-mail john @earthyreport.com or write to Earthy Report, 44 Greenbay Road, Winnetka, IL 60093.

This book was set in Calibri by Amox Enterprises. Printed and bound in Canada.

Library of Congress Cataloging-in-Publication Data

Green Bites: Ecological Musings from the Front/ edited by John Vlahakis.

Includes bibliographical references and index.

ISBN 978-0-615-431079 (pbk)

Environment – Current - 21st century. I. Vlahakis, John

WA670-847

363.7

2011921740

*For Stephanie, Alexandra, Ian, and Zachary*

# Contents

**Preface** ......................................................... 8

**Section 1 – For the Earth** ............................. 11
Chapter 1: Environment ............................... 12
Chapter 2: Waste ............................................ 43
Chapter 3: Wildlife ......................................... 60
Chapter 4: Water ............................................ 72
Chapter 5: Solar .............................................. 84

**Section 2 – For the People** .......................... 95
Chapter 6: Health ........................................... 96
Chapter 7: Lifestyle ...................................... 108
Chapter 8: Consumer ................................... 128
Chapter 9: Green .......................................... 149

**Section 3 – For the Future** ........................ 163
Chapter 10: Energy ...................................... 164
Chapter 11: Regulations ............................. 178
Chapter 12: Transportation ....................... 198
Chapter 13: Innovation ............................... 232

**Footnotes** .................................................. 232

During the winter of 2009, I decided to take some time away from the day to day of operating the company Earth Friendly Products. The aspects of writing a book have always intrigued me and I thought that this time in my life would be a good time to try.

To be honest, the thought of writing on a daily basis seemed a bit daunting at first. I didn't think of myself as a writer or even an adequate one at that; but, it pays to have a writer in the family. In seeking council from my brother in law, Ted Anton, an accomplished professor and writer in his own right, I took his advice to heart. He said to try to write a little each day; the discipline comes from deciding when you like to write and then trying to stick to it. I found that writing in the morning suited my style best.

I had an initial direction for *Green Bites* but, like with all things in life, it's easy to get sidetracked as to the direction in which one is headed. As the book evolved, it went in an entirely different direction from what I had initially anticipated. I found that topical issues regarding all things green interested me more than trying to rehash a personal philosophy of green living; and I realized that most of us who follow the path of green living don't need another how-to guide.

For those of you who are contemplating a sustainable lifestyle and have picked up this book, I hope you find it helpful in learning more about daily green issues and how you can make changes in your life; and please keep in mind that this is just my take on things. You will of course have your own thoughts and may or may not agree with mine; but that's a good thing. The more engaged we are in having an open dialogue about our existence on this planet, the more likely it will be that we can find new ways to improve it; and the more involved we are politically and personally, the better off our planet will be and the more likely we are to have our voices heard. I've always been an advocate of challenging authority and prefer to have a more active citizenry than one that is apathetic. So get engaged with the issues laid out in this book.

*Green Bites* can be read anyway you'd like. By this I mean you can read it in a particular order or just straight through. The green issues covered in this book range from sustainable energy, wind, water, and solar to health, lifestyle, transportation, regulations and even pets—all green-related. The title, *Green Bites*, is just that: bits of green commentary and information. Those who know me personally may get the metaphor, but then again, that would take another book to explain, which I may some day get around to writing. In the meantime, I hope you enjoy this one. If you do and would like to learn more, you can catch me almost daily on my blog, *www.earthyreport.com*; and feel free to post your comments.

At this time I would like to thank the following individuals who played a significant roll in this project: my best friend and life partner, Stephanie Anton Vlahakis, and my children, Alexandra, Ian, and Zachary. Many thanks to the bright and talented Elizabeth Roberts, whose commentary and editing of this personal effort has been invaluable. Thanks also to Val Osakada for providing occasional kicks, to Barry Cain, who listens and comments on an almost daily basis, and to Kelly Sufranski, for her insights. For bringing his wonderful creative aesthetic to the design of this book's cover and interior, a thank you is due to Kris Mox. Lastly, I want to thank my parents, Van and Olga Vlahakis. Without them, none of this would be possible. My Dad deserves additional thanks for getting me started down this road.

Is this beginning to sound like an Oscar speech? I think I hear the band playing...

John Vlahakis

Winter, 2010

# Section One: For the Earth

I think the environment should be put in the category of our national security. Defense of our resources is just as important as defense abroad. Otherwise what is there to defend?

- Robert Redford, *actor, at Yosemite National Park dedication, 1985*

## Chapter One: Environment

"You forget that the fruits belong to all and that the land belongs to no one."

- Jean-Jacques Rousseau, *philosopher*

## Hey America—We're Number Sixty-One[1]

A team of environmental experts at Yale University and Columbia University biannually compose Environmental Performance Index rankings. The EPI ranks 163 countries based on ten indicators of environmental protection including levels of air pollution, marine protection laws, water quality, and rate of planting new trees. Which country do you think ranked number one in environmental protection? Well, it obviously wasn't us. In fact, the United States dropped more than twenty spots to number sixty-three. The last survey the EPI conducted ranked us at number thirty-nine; this year we came in at number sixty-three. For 2010, the top countries surveyed included Iceland, Switzerland, Costa Rica, and Sweden.

The EPI assigned a score for each of the 163 countries based on the above-mentioned criteria; depending on how each country met the criteria, they were assigned a number value. Iceland, for instance, scored a 93.5 out of 100 on the index, while the US scored a 63.5 out of 100. According to Marc Levy, Deputy Director of Columbia University's Center for International Earth Science Information Network, "The US is the only country that has a significant number of people that don't believe in climate change. The central challenge of our time is to help people understand what's happening around them." CIESIN's results indicate that the US is lagging behind European countries in curbing greenhouse gas emissions, air pollution, and climate change. The Europeans have been focusing on reducing waste and getting their citizens to change their behavior so that they use fewer resources. It seems our country is still in denial over the environmental changes occurring around us. The poor effort on our part is now documented by CIESIN[2].

## Migratory Patterns Impacted by Global Warming

One result of global warming that has not garnered much attention is the effect it has on migratory birds. Winter migratory patterns are being affected around the world, in some cases forcing birds to fly further than usual. Due to longer spring and summer seasons, birds that typically fly to Africa now have to fly further south to get into their proper climate (their return trips north consequently take longer). In some instances, birds are flying east to west versus north to south because of global warming; birds in Europe, particularly the Blackcap, now head west instead of south by going from Germany to the UK's milder climate.[3] (Keim)

I've personally noticed the change this season in my own backyard. Mourning Doves tend to come and inhabit our backyard during the spring, so I was a bit surprised when I heard them cooing this past weekend; I found it odd that the birds would show up so soon, as their usual pattern is to fly south to the Yucatan Peninsula and work their way back up north into Canada. This time of year does not provide many feeding options for these birds, at least until late March, so I can only imagine that their ability to survive was impacted. An interesting note, though, is that migratory birds are tending to stay closer to home because of the rise in outdoor bird feeders. Certain species are sticking around in the winter anyway—cardinals in particular—but we're seeing Blue Jays now with greater frequency, which could be due to the proliferation of bird feeders.

Scientists are concerned about the impact global climate change is having on migratory patterns. Nevertheless, new adaptive migratory patterns have begun to take hold despite increasing climate change. The Chiffchaff, which normally migrates to Iberia and North Africa for the winter, now regularly winters in Britain due to the country's increasingly milder areas, especially in the south and west[4]. Eventually, nature seems to find ways to adapt before it's too late—I just hope we can, too.

## Visiting America's Natural Resources

Many of the country's greatest national treasures can be found in our national parks. Parks like Yellowstone, Grand Tetons, Grand Canyon, Yosemite, Smokey Mountains, Florida Everglades and many others attract over fifty million visitors annually. With the onset of summer, more Americans are hitting the road; some of their favorite places to visit are national parks.

One park you must visit, if you have not already done so, is Yellowstone National Park. The grand daddy of all parks in the US, Yellowstone is one of the most intact ecosystems we have; a large variety of mammals, plants and insects call it home. Some of the most stunning scenery found in US resides within Yellowstone's borders; it would take you days to visit and traverse the park's most breathtaking highlights. With the exception of RV traffic, cell phone service and other modern conveniences we've all come to expect, Yellowstone looks a lot like it did when white surveyors happened upon it for the first time. It is a testament to doing what's right to protect our natural resources.

Yellowstone has certainly had its ups and downs, particularly when park rangers hunted the wolf to extinction in the 1920's, only to realize sixty years later that they needed the wolf back in the park to keep elk populations down and maintain the natural fauna. Nature needs it predators to keep ecosystems in balance; our interference by hunting down the wolf ended up devastating the local Aspen tree populations, which fell prey to the elk. By maintaining an equal balance within the park by reintroducing wolf packs, the Aspen trees have made a comeback, and the elk have been kept in check.

I highly recommend that you book a trip to Yellowstone. Please just be sure to leave the RV at home.

## Climate Change Refugees

One little known fact that has begun to make its way into US papers is the growing international crisis of climate change refugees. What are climate change refugees? People who become climate change refugees are those forced to leave their homes due to rising sea levels, droughts and lands that have been consumed by deserts; they are not political refugees motivated to leave their homes due to war or religious persecution. They are leaving their homes because the land they live on can no longer sustain them[5].

This international crisis is most visible in Africa, particularly in the countries surrounding Kenya. There are currently over 250,000 climate change refugees in Kenya; at least ten million refugees exist in other parts of the world. The UN has projected that that number will grow to twenty-five million by the year 2050.

The crisis in Kenya is the result of a four-year-old drought that has not abated from surrounding countries. Refugees are seeking access to countries that have the resources to sustain them— but Kenya is not one of those countries. The country is already experiencing the pressure of new people demanding more of its limited resources. The UN is doing what it can to assist, but this is a global issue that will eventually need to be dealt with on a larger scale. We need to do something now, not later when it's too late[6].

## Geo-Engineering Getting Hotter

Geo-engineering entails using the oceans to absorb more carbon dioxide and adding sulfur to the atmosphere to block the sun's rays; geo-engineering more or less has us manipulating the atmosphere and oceans of the world in an attempt to lower the earth's temperature. Many people are viewing it as a dire response to global warming, as it is by no means a real solution to combating rising temperatures.

Scientists have noted from past volcanic eruptions that the amount of sulfur spewed into the atmosphere has an impact on global temperatures; the last major volcanic eruption—Mount Pinatubo in the Philippines—cooled the planet by 0.9 degrees[7]. The problem with adding sulfur to the atmosphere, besides the affect it has on human and animal lung function, is that it gives us a very hazy sky. This could have an impact on lowering solar energy production and make it difficult for astronomers to view the celestial skies.

Discussions on adding sulfur to the atmosphere have focused on using aircrafts to spray aerosol at high altitudes. There have also been discussions focused on finding ways for oceans to absorb more carbon dioxide; yet, even if the oceans could absorb more carbon dioxide, marine life would still be in danger due to man-made pollution entering the oceans.

Despite scientific investigations on the practical aspects of geo-engineering, the nonprofit organization, the Institute of Physics, indicated that, "Climate geo-engineering at scale must be considered only as a last resort...There should be no lessening of attempts to otherwise correct the harmful impacts of human economies on the Earth's ecology and climate." Many of us would rather see government and industry do everything possible to prevent global geo-engineering. Man's tinkering has never worked out and something of this magnitude should be left alone[8].

## Non-Chemical Fertilizers

Most of us who garden know that fall is an important time to prepare our lawns for the next year's enjoyment; but not as many of us may know that using non-chemical fertilizers can provide for a healthy lawn as well as for some wildlife benefits. Organic fertilizers derived from seaweed, fish emulsion, dried sewage sludge, corn by-products, and poultry and cow manure all provide excellent soil nutrients. These types of organic fertilizers release nitrogen slowly into the soil; the steady release of nitrogen allows for slower grass growth and prevents grass burn, which can occur from the use of chemical fertilizers. Organic fertilizers help rebuild subsoil levels and excel on lawns with very little topsoil.

Besides the use of non-chemical fertilizers, the use of water can be regulated for maximum benefit to your lawn's health. Most local ordinances have a watering ban during the day; but, if you can water in the early morning when humidity is at its highest, your lawn will retain that water for a longer period of time, since morning dew helps retain water better than evening cool downs. Another approach to maintaining a healthy lawn with non-chemical fertilizers is in the way the lawn is cut. Try to cut high and allow the grass clippings to remain on the lawn; the additional clippings help provide for future nitrogen releases. Make sure your lawn has plenty of sunshine and adequate drainage; cut back tree branches to let sunshine in and see to it that water drains easily.

Wildlife begins to come back onto lawns that are not chemically maintained; frogs and toads will become residents, since an organic fertilizer does not impact their ability to draw in oxygen through their skin (chemical fertilizers seep through their skin and can kill them). Now the only issue that remains is who's going to mow the lawn this weekend?

## Are Deserts Shrinking?

As a consequence of global warming, it had been expected that the world's deserts would increase in size; yet, a new prediction came from a group of scientists stating that global warming is actually causing deserts to shrink, primarily the Sahara desert. New satellite images show that the Sahara is not growing, but that new patches of green growth are developing in parts of the desert. It is all due to global warming, according to the scientists. Global warming is also currently increasing rainfall levels and is helping turn the tide of desert growth; the growth and shrinkage of deserts is a little bit more complex than getting a higher yield of rainfall[9].

The overuse of land for agriculture hastens the creation of new deserts; areas in Egypt, and even in Italy, have seen increases in once tillable lands turning into deserts. Ground contamination from pesticides and fertilizers has impacted ground aquifers used in agriculture and created brackish water that cannot support farming. As this process occurs, lands begin to turn into deserts.

The impact from global warming has not fully played out and there are scientific arguments from both sides of the aisle disputing their respective positions on this matter. If deserts that are not used in farming can be reclaimed, though, then this is an unexpected positive outcome of global warming[10].

**No Impact Man**

Have you heard of the Colin Beavan, the No Impact Man? Colin came out with a documentary and a book highlighting his one-year journey of being the "No Impact Man". No Impact Man asked his wife and daughter to forsake appliances and other conveniences they take for granted; conveniences like electricity, cars, trains, planes, buses, eating out, or acting like an ordinary consuming American. They could not use the elevator in any building (and they live in New York City!), and could only buy foods that were in season; almost everything they ate came from a farmer's markets. They also gave up their TV and even had it removed from their apartment; water usage was kept to a minimum and there was no hot water. Everything they did was done to reduce their $CO_2$ emissions to zero and have no impact on the utilization of resources.

One of the more light-hearted moments was his wife's refusal to give up her Starbuck's latte habit. The one exception in No Impact Man's own life was his laptop; the family still connected themselves to the Internet.

In our own daily lives, having zero impact would no doubt be difficult. I'm not sure I could give up all of my creature comforts; but it would be a noble experiment if we all tried it for a year. Who knows, our lives could become that much richer[11].

## Cattle Gas

Human industrial contributions to global warming have been well documented in "countless scientific research studies." Several studies have pointed a finger at the methane gas production of farm animals in this country, as well as other countries, and their impact on global warming; the methane gas being released because of its massive numbers of sheep, for instance, has impacted New Zealand's air quality.

Researchers in Canada have been working on genetically modifying cattle to reduce their cattle's methane output; the Holy Grail in Canadian cattle research is finding a gas-free cow. In all fairness to the Canadians, they're also working on lowering the feed consumption of cows to further assist in the reduction of cow gas; but, if a cow eats less and still produces the same amount of milk and meat, it will also make a contribution to sustainability and farm economics. I can imagine it would smell much better on the farm, too.

Obviously, as the world's population grows and the demand for food products continues, the demand for a gas-free cow that eats less could help provide greater sustenance to the world and lower methane levels. Now, if they could only do something about Uncle Jimmy[12]...

## Earth Day

The first Earth Day celebration occurred in 1972; the celebration galvanized our society into action. Congress created the US Environmental Protection Agency and, along with State and Local governments, actively moved to regulate pollutants and polluters. Since that time we've had various ups and downs with regards to the direction of improving our environment; some administrations pulled back while others moved forward. All in all, though, it seems we are in a better place today than we were back then—but we still have more work to do[13].

Sometimes I get requests to speak at local schools. Most of the kids I speak with tell me how they are raising money for the Amazon rainforest or for some other far away place. I commend them but also recommend that they start making changes at home and in their neighborhoods, too. It's important that we pay attention to our own homes and backyards first. You can start by cleaning out your home of toxic chemicals. Look to find ways to save energy and water, and reuse items in your home. Solar, geothermal and wind power have come more financially accessible; try planning to take yourself off the grid. If we can turn these changes into habits, perhaps some day we won't need an Earth Day to remind us to do the simple things[14].

## Global Warming Impact on Glaciers

Glaciers all across the world are melting at an alarming rate. Global warming, as we know, is viewed as the chief culprit. At the current rate of loss, sea levels across the globe are expected to rise seven meters by the end of this century, meaning that low-level land areas would literally be under the sea. Large numbers of populations would have to be relocated to higher ground, placing greater demands on agriculture and land usage. The melting of glaciers will most surely impact our access to fresh water. Snowfalls in glacier fields have already been diminished and the ability to expand glaciers has ceased to exist.

Some parts of the world, like in Iceland, may experience the short-term benefit of increased hydroelectric power due to increases in water flow from melting glacier fields; short-term benefits from melted glacier fields could provide a new opportunity for agriculture. Still, existing shorelines are already experiencing rising ocean levels. The US is seeking to build up sand dune areas in barrier islands due to the rising levels, as government reports raise the possibility of Florida and Manhattan being underwater as sea levels rise.

**UK Organics**

The London Natural Product Expo is one of my favorite trade shows.  European product design is quite different from what we see in the US; the design element extends into both packaging and labeling.  Besides design, the English and Europeans who exhibit at the London Expo have an extremely high standard for organic products, and their standards extend far beyond food products.  Personal care products, for example, have a significantly higher threshold for calling themselves organic in the UK and Europe vs. here in the US.  The trade groups here in the London Expo push companies that exhibit to develop more organic components and products; sustainability of ingredients is also carefully monitored, as are fair trade practices.  These same issues are beginning to gain a foothold in the US, but we have a long way to go…

## Beach Replenishment Fights a Losing Battle

US beaches, along with many other beaches around the world, are fighting a losing battle against global climate change. Rising sea levels are forcing governments and homeowners to face the prospect of higher costs in protecting their beaches. Past efforts at stemming rising waters included building jetties and sea walls. Today, the preferred method is beach nourishment, a process that pumps sand from the ocean floor back onto the beach; it involves rebuilding sand dunes and restoring shoreline beachfront. The problem with rebuilding dunes, though, is that with each passing year they need to be built higher. As the dunes grow in order to protect homes and land areas, new conflicts arise with homeowners.

In New Jersey, a state that garners over $20 billion in tourism dollars due to its 128 miles of uninterrupted shoreline, beachfront homeowners want compensation for lost ocean views caused by ever-growing dunes. Many beachfront homeowners have already been given compensation; in the small town of Harvey Cedars, NJ, a homeowner was recently awarded $480,000 for lost ocean views. The town had offered $300, but a court ruling overturned it. The whole beach replenishment project is now in jeopardy in that community.

Homeowners who have oceanfront properties do know the risks associated with their land; hurricanes, winter storms and rising ocean waters can and will eventually damage or destroy these oceanfront properties. Expecting local government to compensate for lost views is just plain ridiculous.

**Eastern Forests in Decline**

The American Institute of Biological Sciences publication, *BioScience,* recently presented an article on the decline of forests in the eastern part of the US. Mark Drummond and Thomas Loveland of the US Geological Survey authored the *BioScience* article. As part of the USGS' Land Cover Trends project, they examined changes in the eastern region of the country using remotely sensed imagery, statistical data, field notes and ground photographs. They found a 4.1 percent decline in total forest area—a "substantial and sustained net loss" equivalent to more than 3.7 million hectares. The researchers described considerable regional variation, with net loss being particularly marked in the southeastern plains; the loss occurred even though reforestation of abandoned fields and pastures continues.

Most net forest loss occurs as a result of urban expansion and timber production, which keeps some land free of forest. Mountaintop removal in the Appalachian highlands has also had a "substantial impact" on eastern land cover, contributing to more than 420,000 hectares of net forest decline. Drummond and Loveland indicated that their findings suggest that forest transitions may not plateau or stabilize after reaching a point of maximum recovery. According to the researchers, that "has important implications for sustainability, future carbon sequestration and biodiversity."[15]

**A New Geological Era**

Science has viewed extinction events as a sharp decrease in diversity and abundance of macroscopic life. There have been five extinction events in the Earth's history: Cretaceous, Triassic-Jurassic, Permian, Late Devonian, and Ordovician. According to an article in the journal *Environmental Science & Technology*, we may be about to enter the earth's sixth largest geological era, which could have massive implications for humanity and the planet. If our planet does move into a new geological era, it will be due in large part to the dramatic changes caused by global warming and species extinction. Scientists contend that recent human activity, including sprawling cities, high population growth, increased use of fossil fuels and the extinctions of many plant and animal species, have changed the planet to such an extent that we are about to enter a new era. The exact beginning of this new era, also know as the Anthropocene Epoch, may have already begun when man first began to farm ten thousand years ago or with the increased use of fossil fuels; ten thousand years in geological time is but a small fraction of a geological period that extends into millions of years.

So, are we in a new geological period? Based on past extinction events that took millions of years to occur, it would be doubtful to assume that we are at the beginning of one, especially when taking into account that this is a man-made event. It's giving humanity a lot of credit that as a species we would endure for the millions of years that would be necessary to equate this as a geological event[16]. I'm not sure I'd bet on us lasting that long as a species[17].

## Whole Lot of Shaking Going On

The world has had a lot of shaking going on lately. Is this recent rash of earthquakes and volcanic eruptions normal or is there something more sinister afoot? Has the building of the Ganges Dam, which coincidentally tilted the Earth's axis by one inch, been behind this extra activity? Is the earth finally saying enough is enough? Well, according to researchers at the US Geological Survey, 2010 is really no different than any other year.

Each year we can expect six to thirty-two earthquakes to occur at a magnitude of 7.0 or higher[18]. So far in 2010, we've had seven 7.0 earthquakes throughout the world and are on pace to hit sixteen by the year's end. Volcanic eruptions are not normally factored into the annual earthquake equation, but they are still considered seismic events. The last eruption in Iceland caught the world's attention due to its impact on flight traffic in and out of Europe, though most volcanic eruptions don't impact people to quite the same extent; but the earthquakes that have hit around the world this year have hit heavily populated areas, causing higher levels of destruction and human casualties. The USGS says that the recent earthquakes are no different than any other year except for the amount of destruction they have caused. They also say the 7.0 earthquakes will neither be greater nor lesser in number in the years to come.[19]

Though we don't always know where earthquakes will occur, we do know that we will get our share of them each and every year. That's just the way of the world.

## Oil Drilling Should Pause

What would you do if 100,000 gallons of oil spilled onto your lawn, your street, the local schoolyard, parks, or just about anywhere you walk or drive to?  How would you handle it if those 100,000 gallons turned out to be a daily dump into your neighborhood and the surrounding area?  This isn't a rerun of the Beverly Hillbillies, folks; it's what's going on in the Gulf of Mexico.  It's an ecological disaster that will take years to fix.

The most recent oil spill exceeds the worst-case scenario presented by Beyond Petroleum (previously British Petroleum) when they filed for permission to build an oil platform in the Gulf.  Imagine that: an oil company downplaying the worst-case scenario of an oil spill's environmental impact.  To be fair, it's not just BP's fault.  The agency that regulates such occurrences, the Interior Department's Minerals and Management Service, did not require BP to file a scenario for a potential blowout (referring to the sudden release of oil from a well).  BP's worst-case estimate was 162,000 gallons of oil per day.  As far as scientists can tell, that estimate is proving to be pretty accurate.

What I find amazing is that our own government would even allow BP's worst-case estimate to stand.  Isn't there a better way to contain such an accident?  Are there no better safeguards or containment options when we allow oil rigs to be set up in such deep waters?  The current administration wants to encourage new drilling and is getting Congress to go along with it; but we don't have the means to prevent such ecological disasters.  We need to consider these factors prior to moving full speed ahead to new offshore drilling.[20]

## Possible Aral Sea Restoration

The Aral Sea, located in central Asia, was once part of the former Soviet Union and the world's fourth largest inland sea. In the early 1960's, the Soviets built two dams on the Amu Darya and Syr Darya rivers to divert them for irrigation. The diversion allowed the waters that would have flowed into the Aral Sea to instead be used for farming. Farming cooperatives then exploded in the region and were highly successful. Unfortunately, the manmade diversion created an environmental disaster.

The Aral Sea is a saline sea. Within the last thirty years, the sea has shrunk by over 65,000 square kilometers, losing half its size. The fallout from the loss of this inland sea has been devastating. The remaining waters are so salty that they no longer sustain aquatic life; the areas of the sea that became dry turned into desert. The local climate has shifted, too, with hotter, drier summers and colder, longer winters. Salty soil remains on the exposed lakebed; dust storms now blow up to 75,000 tons of this exposed soil annually, dispersing its salt particles and pesticide residues. The resulting air pollution has caused widespread nutritional and respiratory ailments; crop yields have diminished by the added salinity, even in some of the same fields irrigated with the diverted water.

Now the United Nations is trying to return the sea back to its former self, though scientists expect that it could take thirty years for the process to reverse itself, even if the dams are removed; but, there has been some recovery in certain parts of the sea. The most striking story of recovery takes place in Kazakhstan's North Aral Sea. Improvements in irrigation efficiency and the construction of an eight-mile dyke have raised the water levels back to forty-two meters; they have also increased the surface area of the northern sea by fifty percent and lowered salinity levels to the point where the indigenous fish populations can thrive again.

The Aral Sea is the most striking example of man's folly in altering his environment.  Hopefully the slow process of restoring the Aral Sea becomes a reality.

## A Canadian *Avatar* Scenario?

Movies can be a wonderful escape from the everyday; whether you like drama, biography, comedy, science fiction or even horror movies, they can all provide us with daily inspiration, commentary and parallels to reality. One of the most hyped films from 2009, *Avatar*, has a strong environmental message for moviegoers and may also parallel a real environmental disaster. Some people equate the film with being a recruitment for eco-terrorism, while many religions view it as a recruitment for the establishment of a new deity, Gaia, the Earth Mother. No matter which camp your thinking falls into, environmentalists are playing off the connection between *Avatar* and one particular environmental disaster area: the Alberta tar sands.

Fifty environmental groups launched an advertisement in the *Daily Variety*, a show-business trade publication, characterizing Alberta's tar sands as an ecological disaster comparable to *Avatar* (the ad ran right before the Oscar's award ceremony)[21]. The Alberta Group, which manages and mines the Alberta tar sands, was taken aback by the allegations in the ad, since they considered the similarities to be a stretch; yet, other groups did not and were quick to point out the similarities in landscape and what both corporate groups are trying to mine. The film's corporate group seeks unobtainium, while the Alberta Groups seek to squeeze oil out of tar sands (removing oil from tar sands is a costly and difficult process).

*Avatar* does provide us with a clear environmental message, but I personally felt that it also commented on our past dealings with Native Americans. The movie was entertaining and did what Hollywood can do best: wow the audience. I give two thumbs up to the fifty environmental groups that found an innovative use for a current cultural icon by bringing attention to an environmental issue in Canada[22].

**Twenty Milligrams of CO2**

Twenty milligrams of $CO_2$ will have been released into the atmosphere by the time you have read this sentence; up to one hundred milligrams of $CO_2$ will be released upon completion of this entry. Did you ever think that reading something on the Internet could cause $CO_2$ emissions? Milligrams may not seem like an awful lot of $CO_2$ output, but when you take into account how many pages are read daily on the Internet, the numbers add up quickly. Many of us don't appreciate the fact that by using our computers we actually contribute to global warming; even McAfee, the anti-virus software firm, reports that the amount of electricity used to send out spam emails on an annual basis equals the amount of $CO_2$ emissions that three million cars produce annually[23]. Coal-fired energy plants, which light up our computers, are the main culprits here. Scientists are warning us that if we don't take actions to significantly reduce our emissions within ten years, seismic climate changes will occur.

The greener the power behind the Internet becomes, the better it is for all of us; it's part of raising our sustainability consciousness. Time to get those solar panels in to power the laptop[24].

**Reducing CO2 Emissions**

As many of us already know, the increase in CO2 levels has had an impact on the rise in global temperatures. The issue has prompted governments, through the Kyoto Pact, to set goals to reduce these levels. Our own efforts here have been hampered by politics and an unwillingness to make tough choices in reducing these emissions; yet nature already has an answer to all of this: planting more trees.

We as a species need to address the restoration of forests as an environmental priority. If the planet and humankind are to survive, we need to get these emissions down. The best way to start is by replanting forests. In the Northeastern part of the US, this has already happened. The forests in that area had been cleared three times; but residents realized that the soil was not going to sustain farming. Consequently, they let the forests grow back, which has had an impact.

Ohio, Indiana, Michigan and Wisconsin were once forested lands. I'm not advocating returning these states to forests; but we could stop building strip malls and subdivisions and start building forests in those areas. Instead of worrying about the loss of forests in the Amazon, every country needs to address their own deforestations. Trees are the most effective tools we have in reducing CO2 levels; and we need more of them.

## Hair Conditioners Take on CO2

The American Chemical Society recently reported that amino silicone, an ingredient found in many hair conditioners and fabric softeners, has the potential to reduce $CO_2$ gashouse emissions. The implications for this ingredient could help remove ninety percent of the carbon dioxide emitted from flue gases.

Flue gases, not to be confused with swine flu, are gases emitted from coal-fired power plants—the world's worst emitter of $CO_2$. The gases produce up to 2.8 billion tons of $CO_2$ annually. Globally, there are 50,000 coal-fired power plants in existence, 8,000 of which are located in the US. Amino silicones would act as scrubbers to capture the flue gases, which rise through large chimneys; in other words, amino silicones are what trap carbon dioxide emissions[25].

Amino silicones are known as super softeners and are noted for their ability to impart fibers and textiles with a soft, silky feel; amino silicones in hair conditioners interact with hair fibers to provide added softness. General Electric is conducting testing to see how this material can be incorporated into flue scrubbers, although a timeframe for inclusion of amino silicones into such a process has not been announced.

Who would have guessed that your hair conditioner could have such an impact on reducing $CO_2$ emissions? The question now is, can we add a bottle of hair conditioner to our car's tail pipe[26]?

**America's Antiseptic Wilderness**

The US has a great National Park Service and National Wildlife Refuge System; Americans by the millions visit these parks annually and enjoy what they have to offer. On a recent visit to Yellowstone National Park, I came away not only awed by its magnificence, but also slightly troubled by the way it was presented. Yellowstone is the largest and oldest national park in America. It has a long storied history that dates back to when Lewis and Clark first passed through it; but in some respects, it has become an E-ride in our natural landscape.

Beyond the breathtaking landscapes of Yellowstone, there are four entry points into the park. Once you pay your admission, which I personally think needs to be raised, you travel along a well-kept roadway into and around the park. For those of you who prefer to see new places from a seated position, the road is perfect for your first experience in Yellowstone; you never have to get out of the car to see the beauty of the landscape, nor do you have to get out to see the wildlife. You can keep that 7-Eleven Big Gulp handy to wash down those harrowing encounters with nature—it's a totally safe and antiseptic experience. Even if you stop the car to take a picture of Bambi, the US National Park Service is quick to show up to protect you from her; there is no taking your life into your own hands if you decide to interact with the wildlife.

Once inside the park, you'll most likely head for the usual tourists traps like Old Faithful and the ice cream emporium at Yellowstone lodge (you rarely see people off the beaten path from the tourist highlights); but, the fact that you can easily transverse the park within the comforts of your car takes away from the true experience of Yellowstone. To fully appreciate one of America's greatest treasures, you need to get out of the car and hike a few miles into the wilderness. If you do, it will take your breath away—literally; the elevation is anywhere from 6,200 to 11,000 feet up throughout the park. All jokes aside, Yellowstone is spectacular when you physically get out of your car and interact with it; the reward lasts a lifetime.  Just make sure to bring the bear deterrent spray for that moment when you run into Yogi Bear's cousin.

**New Cloud Formation**

A UK-based organization called the Cloud Appreciation Society recently proposed the listing of a new cloud formation that has begun to present itself across the globe. The new formation is called the Asperatus cloud, which is Latin for roughen or agitate. The organization believes that this type of cloud formation is unusual enough to warrant its own description in science.

I recently came across a photo of an Asperatus cloud formation and had thought that it must have been manipulated using Photoshop; it seemed too unreal to actually be real. As beautiful as these new cloud formations are, I have begun to wonder if they are actually a result of some of the many current climatic shifts taking place. Could weather extremes, especially rising temperatures and droughts, be behind these atmospheric conditions? Though extreme weather is not associated with these clouds, what kinds of atmospheric conditions cause such expansive imagery?

The CAS has a website highlighting all types of cloud formations, including the Asperatus. If you're interested in learning more about the clouds and in seeing some exceptional cloud images, visit *www.cloudappreciationsociety.org*.

## Feeling the Heat

For those of you who are still in denial over global warming, there comes new scientific evidence that continues to highlight the impact it's having on our environment. In the Artic Circle, local fauna is feeling the heat of global warming. Scientists have compared aerial photos taken in the 1940's by the US Navy in their exploration for oil in northern Alaska (the aerial pictures covered over twenty-three million acres). Comparing photos from the 1940's to the present day have provided new evidence that the Artic tundra is greening and becoming shrubbier. Unfortunately, the new evidence extends southward from the Artic tundra, too; satellite remote sensing indicates that the boreal forests just south of the tundra are browning—the result of dry conditions, increased fire activity and insect infestations. The changes to the land cover may exacerbate thawing of permafrost; that in turn could liberate carbon in the form of carbon dioxide and methane from the peat that had previously been locked up in cold storage, all of which would further contribute to global warming.

There are over four million people currently living in the Artic region. Their livelihoods, which subsist on hunting, commercial logging, transportation and infrastructure, are being impacted by these warming trends. Continuing ecosystem transitions are likely to profoundly affect the human and wildlife inhabitants of this region and could quite possibly intensify the already rising temperatures.

Evidence of global warming continues to grow, but we still lack a coordinated worldwide effort to tackle the issue—nationalistic interests will always trump doing what's right for the environment. Until we can find the political backbone to do what's right, we will continue to reap the errors of our ways.

## A Warmer or Colder Winter

We've only been keeping track of daily temperature data since the late 1800's; geologically-speaking, that data is a mere drop in the bucket compared to the age of the earth and its past climatic upheavals. There are plenty of opposing viewpoints on the current climate change debates: conservatives are calling liberals climate change alarmists, liberals are calling conservatives ignorant about all things scientific and the media is staking out its position based on the political leanings of the corporation that owns them.

Given last winter's cold and snowy weather, it's understandable that the political right says global warming is a bunch of hooey. There was certainly more snow in the Northeast this past year, with a lot of areas setting records; and it seemed much colder this winter to many of us here in the Midwest and further East. Here in Chicago, the winter was listed as one of the snowiest on record with below average temperatures. Pundits would argue that more snow in the East and Midwest is the result of increased moisture in the air due to global warming; all of those ice caps and glaciers have been shown to be melting at a higher rate, according to the US Geological Survey in a report they recently issued. The USGS has shown that the melting of glaciers and ice caps has accelerated in the last twenty years, though the phenomena began back in the late 1800's with the advent of the industrialized age[27].

Interestingly, scientists who study the history of the comings and goings of ice ages find that the earth has a past history of global warm ups and cool downs. What's new to the mix in the future history of the planet's temperatures is man's manipulation of them. Increases in $CO_2$ levels have been found in undisturbed samples of sediment drilled out by geologists. The causes are unknown, but we do know that climate swings in the past have deadly and far reaching consequences for life on this planet. Just where our current climatic swing takes us is not entirely known at this time.

## Summertime Weather Getting Hotter

The National Oceanic and Atmospheric Administration released their State of the Climate report for June, 2010. The report revealed that in June of 2010, the temperature in the United States averaged 71.4 degrees Fahrenheit, 2.2 degrees above the long-term average of 1901-2000. Similarly, the average precipitation level for June was 3.33 inches, 0.44 inches above the long-term average. [28]

Above average rainfalls were tallied in Michigan, Illinois, Iowa, Nebraska, Wisconsin, Indiana, Ohio, and Oregon; below average rainfalls were experienced in Maryland and Virginia. New Jersey, Delaware, and Maryland experienced persistent dryness again for the month of June. The three states consequently began experiencing water usage restrictions.

For most of us who live in the Northeast and Midwest, June proved to be a very hot month. The pattern of weather that brought such high temperatures to parts of the US was the result of a stationary high-pressure system that blocked many storms from bringing relief. A layer of high pressure controlled much of the East Coast, bringing an influx of warm air and contributing to record high temperatures. The high pressure also affected the Southeast; south and central regions had their second, fifth, and seventh warmest June on record. The only region that averaged a temperature below normal for June was the Pacific Northwest.

Although the NOAA report only covered climates in the US, the rest of the world is experiencing similar phenomena. The reality is that global warming has been increasing the average temperature of the earth's surface air and oceans since the mid-twentieth century. Averages in temperature will always have some variations, but the overall data continues to show an upward climb in average temperatures. Until we start doing something about global warming, expect our summers to get a little hotter each year.

## The Future of Oil Rigs

There are now some 6,500 offshore oil and gas installations worldwide, about 4,000 of which are in the Gulf of Mexico; the Gulf produces thirty percent of all US oil output, while offshore drilling represents ten percent of the world's oil supply.

From 2005 to 2008, worldwide offshore drilling grew by sixty-seven percent. Brazil has discovered an oil field off their shores that rivals the total oil reserves of Nigeria. The country plans on implementing one of the largest engineering projects to drill for that oil. According to the oil industry, the threat of future spills from an offshore rig is as probable as a plane crashing; they indicate that oil rigs in general are too secure for spills to happen with any predictability. Unfortunately, the British Petroleum spill in the Gulf of Mexico spill has become the country's largest oil disaster[29].

Oil rigs are more or less floating cities that come with high levels of risk not just when it comes to extracting oil but also when it comes to the workers who keep them going; the rigs can run up to the size of two football fields, with over 130 people working them on any given day. Oil rigs can drill up to 30,000 feet, or about six miles into the earth's crust. They are dangerous places to work, but for those who work the rigs, it can be a very lucrative occupation.

The British Petroleum spill shifted our attention to the safety of oil rigs and to the environments in which they operate. President Obama re-issued his moratorium on future drilling in order to review data from the spill and enact new safety measures to prevent spills from happening in the future. The question that needs to be addressed is whether we can ever prevent a disaster like this from happening again; what safety measures need to be put in place to prevent it? Until these questions can be answered, we need to stop future oil rigs from being built and instead find new preventative measures to keep the other 4,000 from becoming the next big spill[30].

## Chapter Two: Waste

"The packaging for a microwavable "microwave" dinner is programmed for a shelf life of maybe six months, a cook time of two minutes and landfill dead-time of centuries."

-David Wann, *author and President of the Sustainable Futures Society*

**Freecycling—The New Recycling?**

Holidays come and go. Presents are opened, gifts are exchanged, and parties are tabled—at least until the next birthday obligation rolls around. All of this gift giving and holiday partying has one severe environmental impact in common: waste. Paper, packaging, plastic cups, and gifts end up being exchanged or returned; other times, gifts get trashed (when was the last time you received a fruitcake?). Most of this waste stream would likely get recycled, but recycling, as we have learned, is not all it's cracked up to be. So, for this new decade, let's start considering Freecycling.

Freecycling is the concept of giving items to people for free instead of throwing them into landfills. One might say that plastic cups are recyclable and should be disposed of after use; yet, they take a tremendous amount of resources and energy to produce, ship and recycle. Instead, buy ceramic cups and keep using them. Once you tire of the cups, give them away to someone else—one man's garbage is another man's treasure. Most items can also be picked up by organizations such as the Salvation Army, Vietnam Veterans, and the American Cancer Society. Think about that couch you had from college and just can't bear to part with. It could be given freely to someone who could really use it.

By beginning to freecycle, we can cut down on how much we need to produce—thus lessening our consumption of materials and energy—and help lower our carbon footprint by lowering our consumption orientation. Freecycling might even have the added benefit of getting your house back in order!

Instead of consuming to enjoy life, perhaps we could take the longer view of enjoying life's experiences instead.

**The Plastiki Expedition**

The Plastiki Expedition is the brainchild of David de Rothschild, an eco-celebrity and scion to the de Rothschild family fortune. He is also a leading environmental advocate for global change. De Rothschild's latest efforts entail changing people's perceptions about alternative uses for plastic bottles. He calls it the Plastiki Expedition: a sixty-foot catamaran built entirely from plastic bottles and containers. The sea-worthy craft is large enough for a crew of six people, though the living quarters are the same size as the interior of a SUV. The goal of the expedition, besides raising awareness on finding new uses for plastic bottles, is to travel twelve thousand nautical miles from San Francisco to Sydney, Australia. The crew expects to complete the trip over a three-week period with stops along the way. Fresh drinking water is to be collected from rainwater and a bicycle on one of the pontoons will provide exercise and act as the boat's electric generator. The crew will consist of four scientists and two sailors.

**Green Landfills**

Landfills have played a role in human habitation since the beginning of time; but we have yet to find a way to get rid of these foul-smelling, unsightly cesspools of human waste. The amount of garbage we produce in this country, when compared to the rest of the world, is staggering. Finding ways to reduce, reuse, and recycle our waste streams would go a long way towards improving our sustainability. Still, we do need landfills, even though many have been closed and left as open spaces. Some landfills have become wildlife preserves or golf courses, but many remain unusable and cause potential health hazards.

The majority of landfills are wide open spaces that lend themselves to becoming solar fields. So a town in New Jersey decided to do just that. The town of Pleasantville, NJ, converted a previous landfill into a green solar field and then turned part of the landfill into a commercial office park. All of the new buildings that were built on the landfill sport solar roofs and are self-reliant on their electrical needs. The town's approach caught the attention of state officials, who are now looking at other landfill sites around New Jersey to convert to sustainable energy fields.

Reusing our closed landfills gives us immediate access to open spaces. Those spaces can easily be converted into new sustainable energy farms, be it solar, wind, or geothermal.

**Pellet Power**

Agricultural production creates excess plant material that cannot be utilized for the kitchen table. The USDA estimates that only thirty percent of all harvests actually make it to the table, leaving an unbelievable waste of seventy percent in production efficiencies, most of which is then discarded; yet, there are solutions to this waste.

Right now in North America there are sixty energy plants that are taking plant waste and converting it into fuel to run their operations. The process begins with removing all water from the plant material. The grinding of plant material then follows this dehydration process. The plant material is turned into a powder and pressurized into pellets. The pellets are then used as a fuel source. A blower is added to the burn process to increase the heat energy being released. There is even a 3,400 square-foot home in Idaho that runs exclusively on plant pellet power and is completely off the power grid. I have not seen the physical equipment set up on the home, but if it were mass-produced, it could easily be another way to help reduce our dependency on foreign oil.

## E-Waste Battle Heats Up

E-Waste is defined as waste generated by consumer electronic products: laptops, desktop computers, cell phones, TVs, stereos, MP3 players, toasters, etc. A common consumer question is what we do with our electronic gadgets once we decide to trash them. One answer to that question comes from the City of New York, which passed legislation requiring electronics manufacturers to take greater ownership when recycling their products. The new ordinance is forcing electronics manufacturers to take responsibility for goods at the end of their lifespan, which shifts costs to the products' producers. The city argues that hazardous metals found in electronic goods are difficult for the public to dispose of and that industry must bear the burden for their products.

Tech industry groups such as the Computer Electronics Association (CEA) and the Information Technology Industry Council (ITI) are seeking an injunction to stop the city's proposed electronics recycling program. The industry groups are contesting the legality of the New York plan, arguing that it's unconstitutional and designed in a way that places "enormous burdens and costs" on manufacturers. In their motion for a preliminary injunction, the CEA and ITI also singled out a rule in which "direct collection" must be provided for all electronics over fifteen pounds. Other industry groups have indicated that they are complying with current state requirements and pointed out that there are now over four hundred recycling centers set up just for electronic goods.

The growing issue of E-Waste has prompted nineteen states to issue their own mandatory recycling laws for electronic goods, with an additional thirty-one states planning to issue similar legislation. E-Waste has become an increasing problem throughout the world. Establishing meaningful recycling laws could facilitate real gains in reusing, recycling, and reducing the amount of hazardous materials in today's electronics[31].

## Pet Waste into Power

Those of us with pets are part of a group that contributes ten million tons of pet waste each year. The majority of that pet waste ends up in landfills or left to decompose in someone's yard. The problem with leaving pet waste in someone's yard is that as the waste decomposes, it releases pathogens into the soil that eventually end up in water sources. Some of us do pick up our pet's waste and flush it down the toilet, in which case the waste is at least treated by a sewage treatment facility; but like most of us, we put it in a plastic bag and toss it in the garbage can headed to the nearest landfill. The problem of pet waste has now prompted several cities to look for alternatives.

San Francisco is one such city that has been seeking a different approach to pet waste. They're looking at converting pet waste into methane and utilizing that methane to powering homes and businesses. The city is also seeking to find ways to lower the amount of garbage being left in landfills. They claim that at least four percent of the garbage they haul into landfills is pet waste.

The city recently awarded a pilot program contract to Norcal Waste to test pet waste for energy. Norcal Waste will establish a special recycled bag program at some of the city's heavily trafficked parks. Converting animal waste to energy isn't something new; many US farmers are converting cow dung into methane to power their farms, and many European nations are doing it on a large scale. Converting pet waste into energy is using a resource that everyone can do. It might just make picking up your pet's poop a little more meaningful (at least that's one way of looking at it)[32].

**Ban the Butt**

Before you get concerned that we're becoming a little too personal here, please know that what I'm really referring to are cigarettes: cigarette butts are bad for the environment. We all know that smoking is bad for you, but did you know that cigarette stubs are harmful to the environment, too? Cigarette stubs carry a very high level of nicotine and can take up to two years to decompose. They are also known to be a pollution hazard. The casual discarding of cigarette butts at the beach or on the ground around town have an impact on water if they come in contact with it (each cigarette butt contains enough nicotine to contaminate thirteen gallons of water). The butts then have an impact on wildlife, and eventually people, as the nicotine seeps into drinking water sources.

How can we ban the butt? Let's start by banning the cigarette. It's a drug and it can kill; even secondhand smoke can kill. Towns across the country have put smoking bans in place inside buildings and just outside of entranceways. All that smoke can't help the air quality around the world. So let's take that next step and ban the butt altogether. Our lungs and water supplies would be very grateful for that type of legislation. Plus, we'd always have great looking beaches and wouldn't have to worry about getting one of those butts stuck between our toes.

**Ban the Foam**

Polystyrene foam, commonly known by Dow Chemical's trade name, styrofoam, needs to be banned. Polystyrene was first discovered by a German scientist in 1839 and then brought to market as a product for consumers in 1937. It is primarily used today for disposable cups and plates as well as for insulation.

The problem with polystyrene is that its only option is to end up in our landfills, as the government does not allow recycled polystyrene to be reused in food packaging. There are several companies that are trying to recycle polystyrene, but they can only compress it and reuse it in insulation products; several communities in California have banned polystyrene products altogether.

The good news is that there are great alternatives to polystyrene packaging, especially in food packaging. Corn-based bioplastics, for instance, are now being used in plates and utensils. Though most egg producers use styrofoam packaging for their eggs, organic egg producers have dropped styrofoam packaging and moved to recycled PET and corn-based plastics instead. Alternative packaging is also showing up at take-out counters across the country. Corporations like McDonald's need to ban styrofoam products and move to corn-based ones. They alone could make a significant dent in reducing our styrofoam waste stream.

We have better alternatives to styrofoam. Now is the time to use them.

## Phone Book Season Adds Up to Waste

It's that time of year again, right before spring training begins, when your local telephone company leaves a five-pound present by your front door: none other than your latest phone book. The ubiquitous phone book, commonly known as the Yellow Pages, has all but outlived its usefulness in our modern times (does anyone under the age of fifty-five still use one?). If you live in an apartment building, you've most likely seen the proliferation of yellow page droppings; possibly thirty or more books sitting at the mailbox area assuming you'll pick one up and lug it to your apartment. These five-pound phone books eventually turn into ten pounds of unusable pulp; local landfills are filled to the brim with them. It makes me wonder just how many of these books are actually being recycled. Some communities don't want you to throw phone books in with the rest of the recycling; others have specific dates for picking up just phone books alone.

I know that phone books generate lots of revenue for phone companies, as well as excitement for people who love to see their name in them; but the wireless world we now live in has made the yellow pages obsolete. Smart phones and the Internet provide you with the listings you're looking for and most have the added benefit of written reviews. Phone books thus become a waste of resources in our now finite world. By no longer using phone books, we can save on the amount of pulp needed to print them and on the cost of fuel to transport them to homes and businesses. Dropping phone books would also lower our greenhouse gas emissions.

Take the time to contact your local telephone company and ask them to come back and pick up their books. If you can't find the number, try looking it up—it's on the third page of your yellow pages.

**Garbage in, Garbage Out**

As we struggle to find new ways to lower our carbon footprint, perhaps we should take a look at the Europeans. European countries are much smaller than our country (which pretty much goes without saying). Since the countries are so compact, it affects the amount of garbage they collect and generate. Europeans live with less garbage pickups than we do here in the States; some of us here even get twice-weekly pickups. Though the extra pickups cost more, that doesn't seem to be enough to discourage wasteful consumption.

Imagine a world where garbage pickups occurred only once every two weeks. Could you live with that? What would you do to lower your garbage input and output? As a country, perhaps we could do something about all the excess packaging we create; for instance, we could use one hundred percent electronic files for papers and periodicals, finally ending the vast paper stream we have all accumulated. Getting a trash compactor would also help lower the amount of waste we toss into landfills; better yet, we all could learn to compost! If we all applied our American ingenuity, we could find ways to lower our garbage output. By lowering the amount of garbage we create, we would be contributing to lowering our carbon footprint and reducing global warming. Just a few easy tweaks in our use of disposables would make a huge difference in our landfills and in the air we breathe.

**Great Pacific Garbage Patch**

Between the coasts of California and Hawaii lies the Great Pacific Garbage Patch, a mound the size of Texas containing floating waste, primarily plastic containers. Styrofoam, fishing wire, plastic bottles, and other plastic products have all converged into this one section of the world. Plastic containers wash up on shores around the world and do so in massive quantities; Japan estimates that over 150,000 tons of plastic wash up on their shores each year. Though ocean currents pile trash in areas all over the world, the Great Pacific Garbage Patch receives the most media attention. The danger that lurks in this patch goes beyond having plastic entanglements with marine life. Scientists have recently discovered that plastic is decomposing at a faster rate in oceans than was once thought. We know that plastics can take generations to decompose in landfills; but in the ocean, the sun's rays and the temperature of the water decompose plastic at much higher rates (water temperatures need to be at eighty-six degrees to begin the decomposition of plastics). As these plastics decompose, they leech Bisphenol A, PS Oligomer, Styrene Monomer, Styrene Dimer, and Styrene Trimer—none of which are actually found in nature—into the ocean. Unfortunately, scientists don't know what the long-term consequences are for people and for marine ecosystems.

The phenomenon of plastic leeching into the oceans has been observed throughout the world and could be impacting marine ecosystems without our knowledge. The leeching of these compounds has also been shown to be in our drinking water, too, which scientists believe is a greater threat since it's coming from our landfills and finding its way into our drinking water.

Unfortunately, the remoteness of this floating garbage patch is hard to round up and recycle; we don't have the resources to begin a massive clean up of worldwide floating garbage. We need stronger maritime laws to prevent the dumping of plastics into our oceans. The fines must be severe and new ways to monitor compliance need to be developed[33].[34]

**Ocean Garbage: Just How Bad Is It?**

Garbage piles the sizes of Texas have recently been found floating in the Pacific Ocean; similar floating piles of human debris have been sighted in the Atlantic. Our oceans have been the easiest place for people to dump their garbage without worrying about local law enforcement. Cruiseliners and freighters are the main culprits, but many countries along the oceans' edges have also allowed waste to be dumped into their waters.

To drive home the point as to just how bad our oceans have become, take into consideration a recent news brief regarding a beached gray whale that died after getting stranded on a West Seattle beach. Normally gray whales have algae in their gut, typical of these bottom-feeding mammals; but the biologists who examined the gray whale found large amounts of garbage in its stomach. The garbage included a pair of sweatpants, a golf ball, twenty plastic bags, small towels, duct tape, and surgical gloves.

We've all seen pictures in the past of mammals that become tangled in garbage and either suffocate or die of starvation because they are unable to feed themselves. It seems that many of us think the vastness of our oceans will hide our waste without anyone noticing; but the fact of the matter is that aquatic life notices, and the fish we depend on for sustenance are at risk. The time has passed for credible maritime laws to be put into effect. Every seafaring country needs to sign an agreement to ban the dumping of garbage into our waterways. We need to follow the old camping adage, "Leave No Trace."[35]

## 400,000 Volunteers Clean Beaches

Every third weekend in September, the Ocean Conservancy sponsors a global trash pick up. In 2009, the Conservancy's International Coastal Cleanup pulled in over 400,000 volunteers to help clean up local beaches. Volunteers from around the world hit the beaches to clean up after a summer season of tourists and locals. Past events have helped pick up over four hundred pounds of waste per mile of shoreline. The top items picked up were cigarette butts, plastic bags, and food wrappers. In 2008, the total weight of the waste collected came out to 6.8 million lbs! With the volunteers' help, that was 6.8 million pounds of garbage kept from getting into oceans all over the world. Every country with oceanfront property had volunteers working key beach areas. Out in the San Francisco Bay Area, over one million plastic bags were picked up from local area beaches.

The Ocean Conservancy continually tracks forty-three specific waste items in their database. The database also provides annual country-specific data as well as state-by-state data. The numbers compiled over time have helped local municipalities target specific waste prevention measures on beach-related waste streams.

In many respects, our oceans are a lifeline for aquatic life; keeping garbage out of the oceans keeps ecosystems healthy. The Ocean Conservancy's efforts are quite noteworthy. They need our support to continue their mission of ocean protection[36].

**Recycled Glass Pavements**

A company out of Wisconsin called Geosystems has created a porous surface pavement made from recycled glass. The new product is called Filterpave. This recycled glass paving system contains seventy to eighty percent recycled glass, twenty to thirty percent granite, and a polymer used to bind the materials. The glass is acquired from recycling facilities and then ground into a smooth pellet (similar to the consistency of sea glass). The system allows for water to be caught and then filtered through the glass into the ground; it acts like a sponge to release the water slowly. The best part of the system, according to the manufacturer, is that it also captures pollutants within the glass and keeps them from entering the soil.

Filterpave does not act as a heat sink, but instead reflects and keeps its surface area cool. During the winter, the surface areas disperse water and prevent accumulation of snow and ice; icing conditions are reduced and the porous nature of the material speeds up the melting process.

Filterpave has strength equal to that of concrete and can be built to accommodate heavy loads. The time to cure it is less than with concrete: only forty-eight hours. One of the best features of this pavement is the range of colors that can be used. They provide for an interesting visual appearance.

Geosystems states that the maintenance costs of Filterpave are lower than concrete and asphalt. They also insist that costs are similar to current market offerings in paving. It's a novel idea for finding an environmentally-sound solution to recycled glass products.

**Nappy Recycling**

The ultimate in nappy recycling can be found in Toronto, Canada (nappy is a British term for diapers). Toronto, it appears, has the best recycling program in the Northern hemisphere. The city calls it their Green Bin Program. Besides the usual bottles, paper and glass recycling, Toronto has elevated the art of taking soiled diapers and turning them into compost; in fact, the city also picks up used sanitary napkins, kitty litter, and plastic newspaper bags filled with "Fido" waste. All of the animal and human waste is then processed and sold back to the public as useable lawn and garden compost; plastic fibers from the waste are separated and then recycled into plastic pellets (no pun intended) that are used to create playgrounds, park benches, and parking lot car stops. It is truly an exemplary program in recycling and one that can and should be replicated globally.

## Recycling Used Gadgets

How would you like to get paid for that old digital camera sitting in your desk drawer?  You wouldn't need to go to Craigslist or eBay to sell it either.  There is a new company that plans to roll out an ATM-style device that pays you for your used cell phones, cameras, and, eventually, laptops and printers.  The device is aptly named the EcoATM.  It allows you to deposit your gizmo into the ATM, which then scans the device electronically and ascertains a secondary market value for your deposit.  Once the ATM assigns a value, you then have the option to accept the offer for payment or have your device returned.  If your device has no value, EcoATM offers you the opportunity to leave it for recycling.  Instead of receiving payment, EcoATM will plant a tree in your name.  Interestingly, the current EcoATM version uses a camera to ascertain the condition of the item you wish to deposit.  Software tied into the camera then decides upon the payment it will offer.  EcoATM plans on making this service available through large box retailers like Best Buy.

One unanswered question I have for this new service is what actually happens to the devices once the ATM accepts them.  Are they really recycled or are they sold into secondary markets around the world?  If new uses are found for these devices, or if in fact the components are recycled, this would be a wonderful service.  I hope they succeed in finding new uses for old technologies[37].

# Chapter Three: Wildlife

"Every creature is better alive than dead, men and moose and pine trees, and he who understands it aright will rather preserve its life than destroy it."

- Henry David Thoreau, *American essayist, poet and philosopher*

# Wildlife

## Nine Species on the Brink of Extinction

The Aughts decade has come and gone, and as we have begun a new decade, the World Wildlife Federation and Eco Worldly have highlighted nine species that could become extinct in this new decade. Both organizations site environmental changes and poaching as the primary reasons. Environmental changes lend themselves to deforestation, rising global temperatures, and lose of habitats. The nine species at risk include the Amur Leopard, Saiga Antelope, Gorillas, Leatherback Turtle, Pere David Deer, Tigers, Golden Toad, Rhinos, and the Baiji Dolphin, which is already thought to be extinct since its last sighting in the wild was in 2002. The Baiji Dolphin's former habitat had been the Yangtze River in China. China has dammed the river and, along with higher levels of pollutants in the river, has most likely caused this species' extinction.

The Amur Leopard and Saiga Antelope are found in Russia. Due to deforestation and poaching, both species are on the brink of extinction; there are only 135 Amur Leopards left, and the Saiga Antelope population has been reduced by ninety-five percent since 1997. Gorillas in Africa are also on the brink; only 250 Cross River Gorillas and 750 Mountain Gorillas remain in the wild. The Leatherback Turtle has been in existence for over 110 million years; but due to global warming, the beaches where Leatherback Turtles lay eggs have become warmer, thus affecting the turtles' ability to lay their eggs. Consequently, their numbers have been in steep decline since the 1990's.

Tigers are another species at risk of extinction; they have already become extinct in southern China. There are only 3,200 tigers left in the wild. The Golden Toad of Costa Rica is now thought to be extinct, since none have been sighted in the wild. The World Wildlife Fund estimates that half the reptilian species around the world are at risk due to global warming and lost of habitats. The Pere David Deer, originally from China, is considered to be extinct in the wild, with only three small herds to be found on nature preserves.  Rhinos in Africa are also at risk; their cousins, the Javan Rhino and Western Black Rhino, are down to sixty animals left in each group[38].

Global warming, deforestation, lost of animal habitat to human development, and poaching have wreaked havoc on these species.  They are the ones with the highest chance of being lost forever during this next decade.  It would be a crime to watch these species become this generation's Dodo bird.

## Queen Bees Need More Lovin'

According to recent studies, honeybee colonies around the US have plummeted by thirty-five percent since 2008; beekeepers have reported up to ninety percent colony losses on their lands. Some suggestions for the causes of this lose range from stress due to environmental changes, to the toll that pesticides are taking on the honeybee population. Cell phone radiation and genetically modified crops have also been suggested as possible causes. Most likely, it is a combination of all these factors. Researchers say that infections by hidden parasites provide a "perfect storm" that could overwhelm the honeybees' defenses[39].

Scientists from the University of Leeds in England are currently investigating possible causes of the widespread increase in honeybee deaths around the world and think they have found a solution: lots of mating with multiple partners. By having a high number of diverse male partners, the queen honeybee could help protect her offspring from disease. The scientists believe that queen bees and their hives need the genetic diversity that indiscriminate copulation can provide. Queen honeybees will typically mate with up to twelve different male partners in a matter of minutes; some will mate with upwards of twenty different partners. The queen bee of the Asian honeybee variety even takes on forty male partners and in one instance was found to have mated with more than a hundred different male partners.

One area the research report did not discuss was the potential lack of interest on the part of the queen bee to seek out multiple partners. One sobering point of the report indicated that if bees disappeared from the surface of the Earth, man would be faced with an ecological and economic disaster. Survival of the honeybee and its ability to pollinate is vital to the protection of our food supply[40].

## Rats to Face Death From Above

Rats, rats, and more rats. How does an island, an island that is considered to be one of the most beautiful on earth, get rid of a non-native rodent problem? Back in 1918, a boat unloaded rats on Lord Howe Island, an island 800 miles off the coast of Australia; the rats have since been credited with killing off five different birds species on the island and are considered the greatest threat to the native wildlife.

The residents on Lord Howe Island are now taking extreme measures to rid themselves of the rat problem. In August of 2012, forty-two tons of rat poison will be dropped from the air onto the island; helicopters will spend several days blanketing the island with the poison. A similar approach to large-scale rat killing has been tried before, but never in such a populated region. Consequently, the helicopters will drop poison over unpopulated areas, while the poison will be dispersed by hand in developed areas. To avoid the poison being digested by birds and other important wildlife, native birds will be captured and caged, along with cats, dogs, cows and chickens, and transported off the island for a hundred-day period; during that time, children will need to be closely monitored by their parents, officials say.

This is quite an audacious plan in eliminating a non-native species; such an operation of removing native species has never been attempted. The concern we see in this attempt is the introduction of poison into the island's ground and water; and exposing people to poison is not prudent. The islanders would be wise to relocate altogether[41].

## UK Seeks to Reintroduce Insects

When most people think of reintroducing protected species back into the wild, they think of wolves, bison, and bears. Instead, the Royal Society for the Protection of Birds (RSPB) in Scotland plans on reintroducing four species of dwindling insects back into the wild: The dark bordered beauty moth, the pine hoverfly, field crickets, and short-haired bumblebees.

Living up to its name, the dark bordered beauty moth is a lovely species with tawny yellow and brown colors on its wings. It is known as being from only three locations in the UK, all of which are unprotected; the Royal Society is working with the Butterfly Conservation to establish a breeding program for their eventual re-release. RSPB is also working with Scottish Natural Heritage to reintroduce the pine hoverfly in 2011; the fly only breeds in rotting hollow tree stumps, which are largely missing in the UK due to forestry practices. In addition, the Royal Society is working on projects to release field crickets into the Surrey and Sussex Heathlands, and return short-haired bumblebees to Kent.

The Royal Society for the Protection of Birds has successfully reintroduced birds in the past, but insect reintroductions are proving to be an entirely different animal (no pun intended.) According to the RSPB, "Conservation is about much more than simply stopping damaging activities to protect what is there. We have a duty to take positive action to restore species that have been lost. We have the ability to enhance and improve our existing habitats and countryside, and we are actively engaged in trying to achieve that." The RSPB plans to use this project as a launching vehicle for their participation in the UN's 2010 International Year of Biodiversity.

The RSPB's conservation efforts are quite notable. I'm not sure I've come across any plans in the US to reintroduce native insects back into the environment; yet, insects play a vital role in our biodiversity and should always be considered in our efforts to protect the environment[42].

## Bringing Back the Dead

It appears that genetic engineering is being utilized to bring rare animals back from the dead. Scientists from the San Diego Zoo and the Scripps Research Institute in La Jolla, California, are taking frozen cells from dead animals, reprogramming them to become sperm and eggs, and then using them to bring endangered species back from the brink of extinction. The team's long-term goal is to coax induced pluripotent stem (iPS) cells into becoming sperm and eggs; they will be creating iPS cells from tissue held by the San Diego Zoo's Frozen Zoo project, which has samples from some 8,400 individuals representing more than 800 species. The sperm and eggs could be used in in-vitro fertilization treatments to add genetic diversity to captive breeding programs.

Part of the process in creating iPS cells comes from using human genes to trick the animal's genes into creating the cells. The process has not been entirely successful, though. It failed in an attempt to reprogram the genes of a northern white rhinoceros, according to The Scripps Institute. While this is all well and good in trying to perpetuate species that are endangered, it does not negate the fact that these species' natural habitats are under tremendous stress due to human activity. Perhaps if we spent more of our efforts in keeping a rich biodiversity, we would not have to play with genetic engineering. Keeping these species alive for captivity really isn't about maintaining a healthy planet; it's more as a way for the San Diego Zoo to keep their main attractions from dying out, and curbing their revenue stream from tourists[43].

## Rapid Rate of Species Lose Growing

The International Union for the Conservation of Nature, IUCN, based in Switzerland, recently released an updated report highlighting the rapid rate of species extinctions around the world. Their authoritative Red List of Threatened Species revealed that 18,351 out of the 55,926 assessed species are at risk of extinction (IUCN's list covers nearly a third of all known plant and animal species)[44].

Launched in 1994, the Red List is compiled by staff members of the IUCN Species Programme and partners with the IUCN Species Survival Commission, BirdLife International, the Center for Applied Biodiversity Science at Conservation International, NatureServe, and the Institute of Zoology at the Zoological Society of London. IUCN has found that twenty-one percent of all known mammals, thirty percent of known amphibians, twelve percent of known birds, twenty-eight percent of reptiles, thirty-seven percent of freshwater fishes, seventy percent of plants, and thirty-five percent of invertebrates assessed are at risk. "This year's IUCN Red List makes for sobering reading," says Craig Hilton-Taylor, manager of the IUCN Red List Unit. "These results are just the tip of the iceberg. We have only managed to assess 47,667 species so far; there are many more millions out there which could be under serious threat." "We do, however, know from experience that conservation action works so let's not wait until it's too late and start saving our species now," urged Hilton-Taylor. The International Union states that conservation efforts have proven time and again that we can save species from extinction.

Efforts around the world have stemmed the loss of plants and animals; but despite these efforts, we are still losing species. A variety of issues have compounded the threat to species, including global warming, loss of habitat through development, and an ever-growing human population that has placed tremendous pressure on other species; but there is still time to save the species that do exist. All that is needed is the political will to find the way and a public outcry to keep their feet to the fire[45].

**Bluefin Tuna Trade Ban Grows**

Bluefin tuna is found throughout the Western and Eastern parts of the Atlantic Ocean, the Mediterranean Sea, and the Gulf of Mexico. The Bluefin Tuna is loved by sushi fans around the world; but it's a fish that is currently being harvested at alarming rates. Thanks to four decades of over-fishing, it has been driven to just three percent of its original population. The species in the greatest danger of being driven to extinction is the Western North Atlantic population of bluefin tuna.

The country of Monaco proposed protecting bluefin tuna by listing them under the Convention on International Trade in Endangered Species (CITES); Mediterranean countries such as Spain, Italy, France, Cyprus, Greece and Malta quickly voiced their opposition. Still, Monaco's proposal slowly grew support from the European Union and the United States[46], as well as from Norway and Kenya; yet that support proved not to be enough for CITES to provide protection for the bluefin tuna (as well as for other marine species in danger of extinction). Fishery interests and economic gain, especially on the part of Japan, ultimately took precedence, and the ban was rejected[47].

Letting a threatened species survive and thrive means that some day fisheries will be allowed to fish again. It's a short-term inconvenience for maintaining the long-term existence of a vital species.

## Gulf Blue Crab Found to be Tainted

After the BP oil spill hit the Gulf of Mexico, researchers from Loyola University in New Orleans, Tulane University, and the University of Southern Mississippi gathered shellfish to study the effects the oil had on them (shellfish are watched closely because they are a large staple of the seafood industry and a primary indicator of the health of the ecosystem.) Weeks prior to capping the BP well, scientists began to find specks of oil in blue crab larvae. The government said that three-quarters of the spilled oil was removed or naturally dissipated from the water; but the crab larvae discovery was an ominous sign that crude oil had already infiltrated the Gulf's vast food web, and could affect it for years to come. According to Bob Thomas, a biologist at Loyola University in New Orleans, "It would suggest the oil has reached a position where it can start moving up the food chain instead of just hanging in the water. Something likely will eat those oiled larvae... and then that animal will be eaten by something bigger and so on. Tiny creatures might take in such low amounts of oil that they could survive", Thomas said, "but those at the top of the chain, such as dolphins and tuna, could get fatal "mega doses." Biologist Harriet Perry from the University of Southern Mississippi Gulf Coast Research Laboratory indicated that this is the first time they have discovered orange droplets of oil in crab larvae, something she has not seen in her forty-two years of studying crabs. Tulane University researchers are investigating whether the droplets contain any of the chemical dispersants used in the Gulf; but so far, according to biologist Caz Taylor, they have not reached any conclusions.

None of the researchers involved wanted to speculate on the size of the contamination in the Gulf, but did say that forty percent of the areas known to contain blue crab have been affected. Unfortunately for us, we don't know how long it will take to clear the food chain—or how much more marine life will perish because of it[48].

## Ocean Acidification

When many of us think about our current climate woes, we think of all the carbon dioxide output in the atmosphere and the threat it represents to people and ecosystems. Not as much attention is being paid to the impact it is having on oceanic ecosystems.

Carbon dioxide in the air doesn't remain afloat indefinitely. Besides circling the globe, it also settles into the earth and its oceans. The impact of carbon dioxide on our oceans is threatening to become a major problem for mankind; it is raising acid levels in the water and has begun to affect marine ecosystems. As carbon dioxide settles into the oceans, it is absorbed by marine life, particularly marine life that grows shells. Scientists have discovered that the shells of marine life have begun to dissolve into the ocean's water system. The dissolved shells contain higher levels of acid, which are a direct result of carbon dioxide absorption. In addition, as shell marine life absorbs the carbon dioxide, it impairs their ability to grow protective shells—and ultimately, to survive. As the problem works its way up the marine food chain, it will also affect the marine life that subsists on these creatures. That chain finally ends with our own harvesting of marine life.

Marine ecosystems now have a new threat to their existence and to ours. Carbon dioxide is no longer just about global warming and clean air[49].

## Oil Spill Hits Largest Pelican Colony

Pelicans are one of the most exotic birds to watch. On land, they look like sages, carefully contemplating their next move and, perhaps, life in general; in flight, they are some of the most graceful aerialists. As the poet Dixon Lanier Merritt wrote, "A wonderful bird is the pelican. His bill can hold more than his belican. He can take in his beak enough food for a week but I'm damned if I see how the helican!" Pelicans capture our attention in flight and on land, especially when we watch them dive bomb the oceans for fish; but, regrettably, they became one of the casualties of the British Petroleum oil spill in the Gulf of Mexico.

The largest pelican nesting area in Louisiana was hit hard by the oil spill. Researchers from the Cornell Lab of Ornithology underscored the US Fish and Wildlife Service tallies of birds impacted by the spill on Raccoon Island in Louisiana. Government tallies stated that sixty-eight birds in the habitat were impacted by oil, while the Cornell group revealed that 400 oiled birds and hundreds of terns in the area were affected. The group also spotted dead birds but did not provide a tally. Birds that showed blotches of oil but were not covered in it were also at risk of dying; a small amount of oil is enough to kill birds since it hampers their ability to regulate their body temperature.

The Raccoon Island Colony was established by the state of Louisiana in the 1980's; the colony is home to over 10,000 birds and has made successful brown pelican restoration efforts that have taken brown pelicans off the endangered species list. The US Fish and Wildlife Service would not enter the island's nesting areas to save the pelicans that had been affected; doing so would have disrupted other birds and could have done more harm than good to the rest of the colony. According to the Service, roughly 3,000 killed or oil-covered birds across the gulf were collected by wildlife agencies since the oil spill began[50],[51].

## Chapter Four: Water

"We never know the worth of water till the well is dry."

- Thomas Fuller, *English clergyman and historian*

## Water Water Everywhere

Water is life. Without fresh water we and countless other species could not exist; yet the amount of fresh water in the world is only 2.5 percent of the total amount of water found on this planet. There was a news report in which China stated that only 49.3 percent of the fresh water in their surrounding lakes, aquifers, and rivers was suitable for human use (up from 48 percent the year before.) The remaining water was too polluted even for industrial use[52].

The US has an enviable fresh water supply compared to other countries. The largest fresh water supply, outside of Arctic ice caps, can be found in the Great Lakes. The Great Lakes represent eighty-four percent of the nation's fresh water supply. Ten percent of the US population lives around the Lakes and thirty percent of our industrial base calls them home. One of the world's largest fresh water aquifers resides in the US and is located under the National Pinelands Forest Preserve in New Jersey. The US also has one of the most extensive river systems in the world; but, despite all of these advantages, determining how drinkable our water is has been a difficult proposition. There are countless articles discussing the aging infrastructure of our drinkable water system[53]. The majority of our country's infrastructure was built in the 20th century and needs to be upgraded in certain areas; and the US population keeps growing, placing further demands on our water supplies. As we all know, there's a lot of water out there. Unfortunately, securing safe water for human consumption amidst our growing demand will continue to be difficult.

## Water Quality

Just how good is the water coming out of your kitchen tap? The majority of water coming out of the tap around the country is treated, though some treatment plants do a better job than others in removing harmful substances.

Past news coverage has highlighted certain parts of the US in which the water coming out of the tap possessed pharmaceuticals; the reported areas included Southern California, Northern New Jersey, Louisville, and Detroit. The pharmaceuticals found in drinking water in these areas included antibiotics, anticonvulsants, mood stabilizers, and sex hormones. According to reports, forty-one million Americans could potentially be impacted by this contamination; but governmental agencies and utilities have countered that the parts per billion, and even per trillion, of these pharmaceuticals fall far below levels that would be considered a medical dose.[54]

The jury is still out regarding the potential long-term health hazards to people living in contaminated areas. Local areas, and the utilities involved, are conducting long-term studies on the impact of pharmaceuticals found in water supplies; in the meantime, though, they are also working to eliminate the water contamination altogether. Unfortunately, most people at some time or another take antibiotics and other drugs, which often get flushed down the toilet.

If you're concerned about the safety of your water, consider adding a water filter for your tap; it's a good defense. If you don't like the taste of your water, try adding a little lemon peel or cucumber slice to help with the flavor.

## Desalination no Panacea

Desalination, the process of making fresh water out of salt water, is being utilized more and more in arid parts of the world that have ocean access; the demand for water in arid and drought environments has made desalination the poster child for fresh water demands.

The process of desalination is not a perfect silver bullet for water needs. The amount of fossil fuel energy just to get one swimming pool-full of fresh water equals 7500 kilowatt hours. Also, the reverse osmosis of salt water can create fresh water containing high levels of boron content, which impacts animal life and agriculture by reducing calcium and carbonate concentrations, making the water acidic enough to damage pipes. The desalination process creates a brackish waste problem, too. The mixture of seawater and fresh water contains high salt content and chemicals derived from the reverse osmosis process; disposal of this waste can have an impact on landfills if it's allowed to seep into ground aquifers or back into maritime environments[55].

Desalination has become a hot topic within the Middle East, as water shortages have become a growing problem. The next war in the Middle East may not be over territory and religion, but over who has access to fresh water[56].

## Greywater/Blackwater

Greywater, and even blackwater, has become a watering resource for agricultural sustenance. The term greywater refers to the utilization of water from washing machines and dishwashers back into the environment (greywater normally contains remnants of phosphate- and boron-free laundry and auto dish detergents). This water is then used to water lawns and plants around homes or commercial institutions. In areas of the US where rainfall is an issue, many homes, businesses, and governments rely on these systems to maintain their plants and gardens. Blackwater, on the other hand, is human and animal wastewater that has been removed through a filter and then turned into greywater; the waste is turned into compost and then used as fertilizer.

Blackwater and greywater usage are on the rise due to the increasing demands for fresh water. As the glaciers continue to disappear and sea levels rise, a greater emphasis will be placed on using these types of water systems in watering our food supplies.

## Ocean Sprawl

Many of us are familiar with the growing concerns about urban sprawl; but now governments around the world are beginning to look also at ocean sprawl. Much like urban sprawl, ocean sprawl refers to the overcrowding of our oceans, an equally concerning problem that hasn't garnered much media attention until recently. The Obama administration has appointed a task force to begin looking into the way we regulate ocean use[57]; the United Nations' Educational, Scientific and Cultural Organization is examining the way countries are using their ocean resources. Increases in farm fishing, shipping lane traffic and the beginnings of ocean energy use have sparked an interest in managing ocean sprawl. UNESCO, along with the federal government, is also trying to determine where offshore wind or wave turbines need to be placed to help create new sustainable energy.

One would never really think of oceans as having a sprawl problem, since all we see from the beach is an endless horizon; but many of us know that in reality, over fifty percent of the Earth's diverse ecosystem lives underwater. UNESCO is concerned by the depletion of wildlife fisheries and the increase of farm fishing around the world. The organization also has concerns about the way goods are transported by ocean freights and their impact on whale migration patterns[58].

The oceans of the world could be the last frontier for man, but they need to be protected. Finding a balance that protects the ocean ecosystem while meeting our needs is not going to be easy.

**Algae Revelation**

Algae is the original green food additive so fondly pushed by the natural health food channel. It's not quite your biblical revelation, but it is fast becoming a source for fuel and food. Algae pilot plants in Nicaragua are producing high yields of cooking oil and food additive byproducts by using it. In Nicaragua, a local algae plant has found the perfect high yielding algae strain that produces low cost sustainable cooking oil for its citizens. Assisted by US investments, the low cost cooking oil is sold by the government to its citizens. The powder left over from the process is then harvested and sold to companies like Cargill and ADM for inclusion in food products sold here in the US.

Algae grows everywhere, but it does particularly well by the equator, hence the startup plant in Nicaragua. It has a low cost to produce and harvest in that kind of environment and is providing new green jobs in Nicaragua. Who would have thought that the green stuff growing in your pool would someday feed and provide fuel for the world?[59]

## Water Conservation

With each new spring, our thoughts inevitably turn to planting and gardening, which as we well know, require water. Water, though, has become the next big deal for the environment. Global climate change is at the top of the list of concerns, but running a close second is our global demand for fresh water[60].

Believe it or not, one of the largest usages of water in the US goes towards landscaping[61]. One of the best ways to conserve water thus begins with your yard. Start by rethinking what you've planted. Are the shrubs, grasses, perennials, and trees all native to your growing region? By planting local fauna you can make a strong contribution to water conservation, requiring less water to maintain your landscaping.

If you have a sprinkler system installed, make sure it is not watering pavement or your home. Also, lessen the frequency of running your sprinkler. One of the best old-time remedies is to install water barrels under your down spouts to collect rainwater—just make sure you keep using the water that collects in them to avoid creating mosquito breeding havens.

By simply addressing our backyard, we can cut down on our own water usage.

## Clean Water in Six Hours

Four million people around the world obtain their drinking water through a method called SODIS—otherwise known as Solar Water Disinfection[62]. SODIS is an initiative from the Swiss Federal Institute of Aquatic Science and Technology to bring clean drinking water to parts of the world that desperately need it. It involves a simple procedure used to disinfect drinking water; all it takes is a PET water bottle. The SODIS method includes removing the label from the PET bottle (any brand will do), filling the bottle with contaminated water, and then exposing it to the sun for six hours. During this time, the sun's UV-radiation kills diarrhea-generating pathogens in the water. The SODIS method helps to prevent diarrhea and is being used in countries like Bolivia, Nicaragua, Nepal, Kenya, and the Democratic Republic of Congo. This disinfection method is urgently needed, as still more than 4,000 children die every day from the consequences of diarrhea.

Solar Water Disinfection can also be used with glass bottles as well as large plastic bags. The water bags are now being utilized in humanitarian efforts to provide water to areas in need (SODIS bags were used during the earthquake rescue efforts in Haiti). One downside to the SODIS method is that it cannot be used with water that is heavily contaminated or that contains high levels of sediment. Fresh drinking water is becoming a crucial issue for much of the world—only three percent of the water found on the earth is fresh water. Forty-six percent of the world's population lacks water pipelines into their homes[63]. Methods like SODIS are working to provide drinking water for those homes. The best thing about SODIS is its simplicity in solving one of the world's most pressing issues.

**Clean Ocean Waters**

In the past twenty years federal, state, and local initiatives have improved the quality of water by our shorelines—gone are the days of medical syringes showing up on the beach and human waste overflowing into our oceans. The past twenty years have also seen the rapid decline of algae blooms in coastal areas. Despite the threat from rising oceans and warmer temperatures, the oceans around our country have gotten cleaner. Part of that credit goes to an organization called The Surfrider Foundation, *www.surfrider.org.*

The Surfrider Foundation's mission statement focuses on clean water, beach access and preservation, as well as protecting special aquatic habitats. The organization works at both the state and local levels and has a website covering every state that has a shoreline. The website also offers in-depth coverage of everything local that pertains to our beaches[64]. They, and we, want you to take an avid interest in the waters near your home, not just for keeping the waters clean, but also for making sure that erosion, pollution, and marine habitat conservation stay at the forefront of keeping these vital natural resources safe.

## Seeding Oceans with Nitrogen

A recent red dust storm swept across Australia and dumped thousands of tons of soil into the Tasman Sea (located between Australia and New Zealand). The storm at its peak carried 140,000 tons of red soil from central Australia into New South Wales and the Tasman Sea each hour; Sydney itself received 4,000 tons of red dust. The infusion of red dust into the surrounding waters ignited an explosion of microscopic life. Scientists who have been studying this phenomenon indicated that the addition of red dust into the ocean has resulted in an increase in plankton growth (plankton is a key food group for many of the ocean's fish). The increase in growth has validated plans to seed oceans with nitrogen to increase fish yields. The findings of the scientists' research could go a long way in helping feed the ever-growing world population.

The Ocean Technology Group at the University of Sydney argues that the influx of nitrogen into the world's oceans could result in curbing global climate change. This would be an additional benefit to raising the plankton levels from the added nitrogen in the waters. Researchers at the Ocean Technology Group estimated that the growth of plankton within the Sydney Harbor, and up to ten kilometers off shore, tripled in growth once the storm ended—all within a two-week period following the storm. Other parts of the world should try to replicate this natural feat. If indeed it could be replicated, it would go a long way to restoring the balance of the ocean's food chain (particularly with the ability to fuel plankton growth). Nature is definitely lending a helping hand here in showing us a way to help restore the oceans of the world[65].

## Phytoplankton in Full Retreat

The smallest and most important plant to the oceanic world is phytoplankton, an organism that floats on or near the surface of water. It is the foundation of the ocean food chain. The word plankton is derived from the Greek word "planktos" which means drifting; and like the current climate bill, plankton is drifting into extinction.

Scientists have been monitoring phytoplankton for over 100 years and the data stream is that long. Since 1979, the scientists have been monitoring the ebbs and flows of the plankton with remote sensing satellites. Their studies have provided evidence that phytoplankton biomass is fluctuating on a decadal scale and that those fluctuations are headed in a downward spiral; the organism seems to have been disappearing at a rate of one percent each year for the last century.

Since 1950, scientific research has shown that phytoplankton shrunk by some forty percent, according to a report published by the journal *Nature*. Marine scientists David Siegel, from UC Santa Barbara, and Daniel Boyce, from Dalhousie University in Canada, jointly authored the report. Siegel and Boyce noted that the global decline of phytoplankton, which was observed in eight of ten ocean basins, corresponded with a rise in ocean temperatures. The scientists suspect that warming near the surface of the ocean makes each ocean layer more distinct, preventing the bottom layer, which is rich in nutrients, from mixing effectively with the upper layers, thus fertilizing the phytoplankton[66].

Phytoplankton is an important key to the whole oceanic ecosystem. The effect on declining fisheries has yet to be gauged, but in terms of being a harbinger for climate change, it is unquestionable.

## Chapter Five: Solar

"I'd put my money on the sun and solar energy. What a source of power! I hope we don't have to wait till oil and coal run out before we tackle it."

- Thomas Edison, *American Inventor*

## Solar Concentrators to Cool

New solar concentrators have been developed to help cool buildings during times of high demand. The first of these solar concentrators, which will be used specifically in commercial buildings, is being tested by the Southern California Gas Company[67].

Solar concentrators work by reflecting and concentrating incoming sunlight onto a pipe to heat water. The heated water is then used in place of gas or electricity to power an absorption chiller. The absorption chiller is used to create cold air using a heat exchanger and compressor.

According to SCGC, solar concentrators are more efficient and take up less space than photovoltaic panels. Using the sun for cooling has been considered an excellent application for solar energy due to its electrical and gas demands for air conditioning. The concentrators being installed by SCGC are small enough to fit on the roof of an office building and are modular in design; each unit has the capacity to cool down three average-size homes.

There are currently three companies making solar concentrators available to the public: Sopogy, HelioDynamics and Chromasun. There is one known private company using solar concentrators to cool and dehumidify their factory building: Steinway Pianos of Long Island City, NY[68].

**Solar Hot Water**

Hot water is a luxury that many of us couldn't imagine living without; it's something that has been taken for granted by all of us. Throughout the ages, man has used the sun to heat water. Using wooden barrels to collect and heat water continues to be a technique utilized in many countries.

While traveling recently through Athens, Greece, I came upon a sight I had never seen before. It seemed that every home had a solar hot water tank on the roof; not only single-family homes, but high rise apartment buildings, too. It was just like the days of old when TV antennas dotted the landscape.

Greece is fortunate to receive an abundance of year-round sunshine, which people seem to take advantage of on a massive scale. In Athens, I saw nothing but solar hot water tanks. The southern regions of the United States receive lots of year-round sunshine and could benefit from solar tanks, too. Governments at every level should be mandating the installation of such systems. Through tax incentives or direct support, we could alter the solar landscape in this country overnight. It's an incremental way of quickly creating another sustainable energy source.

## Paper Solar Panels

How cool would it be if you could print solar panels from an ink jet printer? Paper solar cells anyone? Researchers at the Massachusetts Institute of Technology just may have the solution to make printing solar panels a reality. MIT scientists have successfully coated paper with a solar cell, part of research projects aimed at energy breakthroughs. The process involves organic semiconductor material that can be applied through an inkjet printer (the coating is similar to the ink used in an inkjet printer). It lowers the weight of future solar panel designs and could potentially be stapled onto surfaces.

The commercial application for such imprinted panels is quite extensive; imagine solar panels on sails, tent fabrics, exterior and interior walls, painted car surfaces, and even clothing. Reducing the weight of solar panels through this new process could eliminate the exoskeleton of solar panel farms and possibly reduce the maintenance costs of current solar panel installations. The materials MIT researchers used are carbon-based dyes. The cells are about 1.5 to 2 percent efficient at converting sunlight to electricity. The size of the individual solar dots being applied to the paper is measured in a few nanometers (compare that to human hair, which is about 50,000 to 100,000 nanometers thick). According to MIT, if 0.3 percent of the US was covered with photovoltaics with ten percent efficiency, solar power could produce three times the country's needs. The scientists have been working on finding the right combination of materials and sizes that can fine-tune the colors of light so that these nanometer dots can absorb sunlight and become candidates for solar cells.

MIT has been quick to caution that these results are several years out in becoming commercially viable products. Nonetheless, their breakthrough has extraordinary consequences for the rest of us, and is headed in the right direction to move us to a more sustainable energy solution for this country[69].

## Are You Ready to Fly Solar?

Would you ever consider flying in a solar-powered airplane? Not a hybrid airplane or one that runs on biofuels, but a craft solely powered by the sun. Imagine a plane that could stay aloft during the day and then fly at night on the energy stored in its batteries. Science fiction, you ask? Not for Bertand Piccard and Andre Borschberg, co-founders of the Solar Impulse. They assembled a team of seventy talented individuals to build the plane. According to Borschberg, everyone on the team was passionate about making a statement about our global dependence on fossil fuels and the untapped potential of burgeoning green technologies.

After seven years of designing and testing the solar plane, aptly named the Solar Impulse HB-SIA, Borschberg took on this maiden flight. The Solar Impulse had already been taking short hop flights across the tarmac at a Swiss military base since November 2009 (the Swiss Air Force allowed them to develop and build the plane on one of their bases). The co-founders' goal was for a perpetual flight at 9,000 feet; in actuality, the plane reached an altitude of over 28,000 feet above sea level[70]. The two Swiss pioneers plan to have their solar-powered plane ready for a flight around the world by 2012.

The Solar Impulse HB-SIA plane is comprised of a carbon skeleton covered in flexible polycarbonate skin. At 3,500 lbs, the plane weighs only as much as a small car; each wing is so light that a single person could carry one. The plane is built for one person to fly and the parachute for the pilot weighs more than the cockpit nose. The plane is powered by a small motorcycle engine and achieved a speed of up to seventy-eight mph. Each leg of Solar Impulse's transcontinental flight is expected to take five days and five nights—slow as molasses when compared to today's jet travel. While it's a bright idea, the practicality of flying a plane at seventy-eight mph would be far too slow for today's air travel. Like all new things, though, this is a first step towards using solar to power a plane; besides, the first flight in 1903 wasn't that much faster. Give it time—we may someday be flying at close to the speed of light[71].

**Solar Lamps to Kill Kerosene**

In Africa, solar lamps have begun to replace kerosene lamps. The move to low-cost solar lamps has helped poor people throughout the continent increase their income levels and improve their health and safety. The toxic fumes from kerosene lamps kill 1.5 million women and children a year, according to UN officials; there have also been cases in which people were burned while using the lamp. The cost of kerosene in African slums can cost as much as twenty-five percent of a monthly salary, while replacing a broken kerosene lamp glass can cost ten percent of one's annual income. The move to solar lamps runs up to fifty percent of an average worker's income, but aid groups have been supplementing the cost to lower its impact on families.

A typical solar lamp can provide a higher output of light than a kerosene lamp, though it only lasts for around four hours before it needs to be recharged, which can take up to eight hours per lamp. One of the cheapest solar lamps made in the US comes from a company called D Light. D Light states that a fully-charged lamp provides eight hours of light on a low setting, which is great for walking outside or socializing, and four hours of light on the high setting, which is intended for any activity that requires bright light. The company also states that the lantern is at least four times brighter than a kerosene lantern, so users aren't giving up lighting quality for off-grid charging capabilities[72].

D Light hopes that their product will become the kerosene lamp killer; I sincerely hope that they're right. The need to help impoverished nations move away from fossil fuels and embrace off-grid technologies is more important than ever. Now all we need is someone to get all of our camping stores to stop carrying kerosene lamps and start pushing solar lamps.

## First Solar Utility Plant Opens

The first national all-solar power plant recently opened in Blythe, California. Operated by First Solar, a solar module maker, the solar power plant generates twenty-one megawatts of electricity, which is enough to power 17,000 homes in Southern California. The electricity generated by this utility power plant is being purchased by Pacific Gas & Electric, who signed a twenty-year purchase power deal with First Solar. First Solar plans to build a number of projects in the area. Recently approved by State regulators, First Solar plans to build a new forty-eight megawatt utility power plant. Pacific Gas & Electric also signed an additional twenty-year purchase power deal for the new facility.

First Solar's initial power plant was solely developed, financed, and built by the company. They built their solar power plant with thin film solar panels made from cadmium telluride, considered to be the most cost-effective panels out there. The lower cost of the panels has made the economics of supplying solar-generated electricity financially viable for utility companies. The new plant also meets the California renewable energy mandate. Private initiative and investing seems to be leading the way to a greener, renewable American future[73].

## Solar Telephone Poles

Commercial and governmental efforts in solar energy have focused on the usage of large tracts of land to build tomorrow's sustainable energy sources; private home use has focused on placing solar panels on the roof of homes. One innovative approach to solar placement has come from the Public Service Energy Group (PSEG), the main power supplier for the state of New Jersey. PSEG recently announced a plan to place solar panels on every utility pole they operate in the state; this will put sustainable solar power in every town. The panels themselves are being called smart panels, due to PSEG's ability to regulate them remotely. According to the group, it will be the largest utility pole placement of solar panels in the world. The placement of these panels on every utility pole is expected to double the solar output that the state currently produces (the projected output of energy is eighty megawatts). This increase will make New Jersey the second largest solar producer in the country after the state of California—not bad for one of the smallest states in the Union.

Public Service Energy Group's efforts are one of the most creative approaches to harnessing solar energy I have seen to date. They maximize the available footprint of an existing infrastructure, which would be great for areas that don't have open land areas. The possibilities of where you can place a vertical solar pole are endless[74].

**Military Solar Power**

The US military has begun to secure solar power fields to provide sustainable energy to their bases. Their first installation is at Ft. Irwin in California's Mojave Desert; it will be the largest military solar field in the world. The 500 megawatt facility will use both photovoltaic panels and solar concentrators to generate power. There will be a second phase of cells added to the installation within the year, which is expected to bring the total megawatt count to 1250 gigawatt hours annually. This type of output has the capacity to power 100,000 homes each year.

The military has also begun to analyze additional facilities where it can utilize sustainable energy fields, since their main objective is to provide a secure energy source for their operations. They have also begun to examine their carbon footprint in terms of security: how viable is alternative energy to military operations and how secure can they make it for their needs?

Leave it to the military to view sustainable energy from a security standpoint. We, on the other hand, view it as a sustainable and environmentally responsible imperative. Nevertheless, I'll accept anything the military does from a sustainability standpoint as a good thing[75].

**Nano Solar Panels**

Solar panel technology is now on its third generation of development. Nanosolar a California-based company, raised significant funding for a new solar panel technology. The new panels are printed through a press, similar to one that prints on paper; they're thinner, more flexible, and lighter than traditional solar panels, but more importantly, they're cheaper.

The old way of producing solar panels involved baking silicon in the panels. The way to do that was in small batches, limiting the efficiency in production and cost. The differences in the old and new way of creating solar panels lie in the type of materials utilized to create thinner panels. The technologies used are varied and range from high-efficiency, low-cost copper-indium-gallium-selenide to cadmium telluride and amorphous silicon. There are about six companies including Nanosolar currently pursuing this new technology. Nanosolar is now shipping panels on a limited basis with the balance of the companies planning 2010 shipments in volume. It seems that things are getting brighter every day[76].

**Old Idea in Solar**

Do you remember as a kid taking a magnifying glass and using it to burn a hole in objects like paper, wood, and possibly the occasional insect? This old custom, which dates all the way back to the time of the ancient Greeks, is now taking place in the fields of Israel. An Israeli Solar company has been building prototype solar fields using the magnifying glass approach behind their solar reflectors. The new reflectors channel light through a large magnifying glass that amplifies the light source within the panels. The energy being cultivated matches current panel outputs but at a much lower cost, according to their estimates (the reflectors themselves purportedly cost less to the manufacturer)[77].

The solar fields look a bit different from the flat panel fields currently in this country; but the new approach is something to consider for our own use as well. As of right now, there is no US distributor selling this approach in our country. As the level of entrepreneurship in alternative energy solutions continues to increase, this idea could become an opportunity for other green entrepreneurs, too.

# Section Two: For the People

"Our greatest happiness does not depend on the condition of life in which chance has placed us, but is always the result of good conscience, good health, occupation, and freedom in all just pursuits."

Thomas Jefferson, *3rd President of the United States*

## Chapter Six: Health

"It is the safest of times, it is the riskiest of times....
What the Dickens is going on here?"

- Denton Morrison, *professor and author*

## Sewage Sludge Hits the Fan

San Francisco recently offered its residents free compost for their gardens. The compost was labeled as "organic biosolids compost" and given away at local community events. In actuality, the compost contained sewage sludge from nine California counties. Once residents discovered the ruse, they promptly returned the "compost" to the mayor's office in San Francisco.

Sewage sludge is the end product in the treatment process of human waste, hospital waste, and storm water. The end product is known as effluent, and is normally incinerated, dumped into landfills, or used to produce methane gas; in San Francisco's case, the city also dumps it into the Pacific Ocean. The problem with all of this effluent is that once the water is removed from it, all of the impurities and toxins are left behind; the things left behind in effluent can include a number of heavy metals, polycyclic aromatic hydrocarbons, pharmaceuticals, steroids, flame-retardants, bacteria (including antibiotic-resistant bacteria), fungi, parasites and viruses. The EPA only requires that communities kill off fecal coli forms in the sludge and ensure that nine heavy metals including arsenic, cadmium, chromium, copper, lead, mercury, molybdenum, selenium and zinc are not present in unacceptable levels[78].

The use and disposal of sludge is a national issue that dates back to the 1972 Clean Water Act. Using sludge in farming applications exists, but the implications don't bode well for farmers or for those who consume the food products from those farms. According to the EPA, about half of all sewage sludge is applied to farmland as fertilizer. Unfortunately, that has lead to instances in which cows have died and farmers have come down with a variety of health issues. The toxins and impurities also eventually enter humans through the food products they eat from those farms. In my view, sewage sludge should only be applied to create methane gas for energy, as continued use of it will cause harm to humans and livestock.

## Food Hysteria

Recently, our morning newspaper had an article about the riskiest foods found in the marketplace—it made front page headlines. It seems that an organization called the Center for Science in the Public Interest published a study highlighting the riskiest foods in the country. The organization compiled data over a twenty-year period and found that one of the most risky food groups is leafy green vegetables: items like spinach and lettuce. During the twenty-year period, the study found that 13,568 people became sick from eating these leafy greens. That's 678.4 people per year out of an average population of 200 million! Let's try to put this in perspective: average 200 million people times three meals a day times 7300 days (twenty years), which equals 438,000,000,000 potential opportunities to get sick from eating spinach and lettuce. What are the odds? Why would the news media even bother scaring people? Granted, we don't eat spinach and lettuce morning, noon, and night; but come on now.

Besides the hazardous leafy green vegetables, the CSPI's study noted that there were also 11,163 people who became ill from eating eggs. The organization's study stated that both food groups could potentially be contaminated with E. Coli, Salmonella, Norovirus and other deadly pathogens; the study further stated that the FDA is not competent when it comes to protecting our food supply. Who is behind the CSPI anyway? Can they be serious? Yes, you can get sick from eating spoiled and contaminated food, but the odds are very slim. How many of us purposely eat raw eggs? Do you not clean your lettuce and spinach prior to cooking and eating it? I'm amazed the organization even got press coverage; are we that hard up for bad news? I find this kind of coverage irresponsible and misleading. Hopefully, people will read the full content of these articles and come to a similar conclusion: eating spinach is good for you[79].

**Diet Can Save Your Skin from UV Rays**

There was a recent report highlighting the idea that a certain diet can save your skin from harmful UV rays. It is the Mediterranean diet. According to Dr. Niva Shapira, a researcher at Tel Aviv University's (TAU) School of Health Professions, the best prescription is to "go Greek." The Greek diet includes foods like olive oil, fish, yogurt, and colorful fruits and vegetables. These foods are rich in antioxidants and omega-3 fatty acids—common in Mediterranean regions—and can protect the skin from the sun's rays. Some of the most helpful antioxidants are carotenoids, which are colorful fruit and vegetable pigments. This includes the reds of tomatoes and strawberries as well as the bright oranges of carrots and pumpkins. Other good foods include whole grains. Foods to avoid include red meat, processed foods, and alcohol with the exception of red wine, which is actually good for your skin. People should also avoid foods containing the compound psoralen (e.g. parsley, celery, dill, cilantro, and figs).

The body's natural defense to protect itself from the sun's rays is to create a skin pigment called melanin. Eating well helps build up your body's ability to create melanin, which allows your skin to tan without harming it. Melanin combines with oxygen (oxidizes), thus creating the tan color in the skin; however, overexposure can cause melanoma, which is a less common type of skin cancer but one that results in seventy-five percent of all skin cancer-related deaths. UV radiation also attacks the immune system, making it more difficult for the body to repair itself. Besides eating well, the researchers commented that one must still use traditional methods such as suntan lotions and appropriate body coverings to combat the sun's rays[80].

## Fear Mongering for Swimmers

The University of Miami issued a report highlighting the potential dangers of swimming in tropical and sub-tropical oceans. Just when you thought it was safe to return to the beach, the university decides to rain on your parade. UM's report is called BEACHES, which stands for Beach Environmental Assessment and Characterization Human Exposure Study (who comes up with these acronyms?). The researchers for the BEACHES report enlisted 1,300 South Florida beachgoers to participate in the study. The group was split between those people who go in the water and those who do not. Several days after the experiment, the researchers followed up with both the wet and dry groups to check on their health status. What they discovered was that beachgoers who went in the water had a slight chance of getting ill from ocean-based microbes. The microbes were not microbes found in sewage runoff, tar balls, or any other type of contamination; they were just plain old natural microbes that live in the ocean. The report also found that those beachgoers who went in the ocean were 1.76 times more likely to develop some form of gastrointestinal sickness and 4.46 times more likely to develop a fever or respiratory illness; plus, the wet group was six times more likely to report a skin illness.

The University of Miami researchers had a few common sense recommendations for us beachgoers: avoid getting sea water in your mouth or swimming with open wounds, shower after swimming, and wash your hands before you eat. Normally, most of us who go swimming in the ocean tend to be more concerned about potential jellyfish stings or stepping on something sharp[81].

The reality is that most people follow safety precautions. I for one do not like getting mouthfuls of sea water and try to avoid it; and I wouldn't want anyone not to enjoy the beach because of these kinds of reports. Use your head when you go to the ocean—and if you hear the Jaws soundtrack playing in the background, head for higher ground!

## Antibacterial Ingredient to be Reviewed by FDA

Antibacterial soap dispensers have become a common fixture in many public areas including airports, grocery stores, schools, and health facilities; the last seasonal scare over the H1N1 virus provided the impetus for the proliferation of these devices.

The majority of antibacterial soaps contain an ingredient called triclosan, which is the primary component found in soaps that make antibacterial claims. Besides being found in antibacterial soaps, triclosan is also utilized in a wide range of products such as socks, toys, and even food items. The Food and Drug Administration recently announced that it would review the safety of the ingredient in antibacterial products. The administration's decision to review triclosan comes on the heels of recent studies suggesting that it can alter hormone levels in animals; past studies have suggested that triclosan increases bacterial resistance to antibiotics. The FDA further stated that there is no evidence that soap with triclosan is superior to soap without the ingredient[82].

Personally, I've never been a fan of antibacterial soaps or products that contain antibacterial properties. My feeling is that regular bar soap and hot water are sufficient in keeping germs off. Antibacterial soaps kill not only bad germs, but also the good germs that combat them; the rise in use of antibiotics and antibacterial soaps is the reason for their growing ineffectiveness in killing bad bacteria and fending off viruses. We've been over-prescribing antibiotics and overusing antibacterial products in our daily lives. All of this overuse has allowed bacteria to become more resistant to the drugs we have. At the very least, we should follow the European Union's ban of triclosan in products that come into contact with food.

## New Learning Keeps Brains Healthy

Ever notice how at least one of your parents picks up the game of bridge in their later years; they find a local bridge group to join, learn the game, and all of a sudden feel mentally alert again? Now there is scientific evidence to show that as we age we can hang onto our cognitive abilities and memories by learning new things. According to UC Irvine neurobiologists Lulu Chen and Christine Gall, everyday forms of learning animate neuron receptors that help keep brain cells functioning at optimum levels. These receptors become activated by a protein called brain-derived neurotrophic factor, which facilitates the growth and differentiation of the connections (or synapses) responsible for communication among neurons. Brain-derived neurotrophic factor, BDNF, is key in the formation of memories. Chen and Galls' findings confirm a critical relationship between learning and brain growth.

In addition to discovering that brain activity sets off BDNF signaling at sites where neurons develop synapses, researchers have determined that this process is linked to learning-related brain rhythms called theta rhythms, which are vital to the encoding of new memories; these relationships provide the necessary evidence for maintaining good brain health. The researchers point to evidence that theta rhythms weaken as we age and suggest that this can result in memory impairment; on the other hand, they suggest that staying mentally active as we age can keep neuronal BDNF signaling at a constant rate, which may in fact limit memory and cognitive decline. Researchers are now exploring whether learning-induced growth signals decrease with age and, if so, whether this decrease can be reversed with a new family of experimental drugs[83].

Even if you think you know it all, you are never too old to learn something new. Consider the possibilities of taking up a new hobby or a class in something you always wanted to learn; traveling, volunteering, or even learning a new instrument or language can keep your brain healthy and young at any age.

## Inoculations are Debatable

During this past year, our government actively immunized those in the population they considered to be most at risk of contracting the flu virus; children and the elderly were immunized for the HiN1 virus and the regular flu virus. Fortunately, the fear of a H1N1 pandemic never came about. According to government scientists and the Center for Disease Control, the fact that we inoculated children first was most likely what protected adults in the general population.

Every year we are placed on high alert for the next pandemic, ever fearful of the next one that will strike. I recently had my yearly physical and discussed flu shots with my doctor. He asked me why I didn't get one. I said I have had the privilege of being exposed to flu viruses for some fifty years and personally don't feel that I'm in a high-risk category. My feeling is that the body's defenses need to be able to combat whatever viruses are out there. I've also heard of plenty of people who get their flu shots regularly but still get sick; so I take my chances.

My doctor's response to my comments was not what I expected to hear from someone in the medical community. As it turns out, many doctors, including my own, are beginning to argue that our efforts to immunize people are actually weakening us; in a sense, we are reducing humanity's ability to fend off future pandemics. Imagine if the mumps mutated into a new virus form. Only those who had previously had it would have the antibodies built into their systems to fend it off.

The most sobering note my doctor put forth was his belief that there would eventually be another pandemic. Our current health infrastructure can only handle five percent of the population for respiratory illnesses; it would be overwhelmed by a pandemic. Doctors and scientists feel that pandemics and viruses like the flu are part of nature's way of making humanity stronger—killing off the weak so that the strong may survive. It's something to consider the next time the government seeks to scare us into getting next year's flu shot[84].

## Health Supplements

Did you know that the natural health food supplement industry is the largest component in natural health food sales? Even conventional drug stores now have large sets of health supplements. One such supplement, glucosamine, is widely used for osteoarthritis—knee osteoarthritis in particular. Most physicians will tell you to take glucosamine for three months and see if there is any improvement. I've used it in the past, but after some time did not notice a difference; yet I learned that those who take it as part of their supplement regime are actually raising their risk for diabetes—especially the elderly.

The FDA is not regulating glucosamine or the majority of other products in the marketplace. Although clinical studies have been done on glucosamine, as well as on other supplements, the time is right for the FDA to start examining the supplements we take. At the very least, it would provide the necessary warnings to allow consumers to make informed choices. Like all consumer channels, the likelihood of manufacturers and marketers governing themselves is slim; only when the federal government decides to step in do manufacturers take notice of the potential new regulations they will face[85].

Being an informed consumer is always your best defense. The web now provides us with the tools to do so.

**A Natural Remedy**

Just in time for summer, along comes a natural remedy for a variety of ailments: vodka. As it turns out, vodka serves many purposes. Pouring it into a full watermelon was always a fun pastime; but learning that vodka's a great counter measure to the effects of poison ivy was a new one on me. Having issues with bug bites? The next time you go out on a picnic, don't grab that can of insect repellent containing DEET—grab some vodka instead. Dab a bit of it on a soft cloth and lather yourself up. It will keep the bugs from biting (though you may run into an issue with local law enforcement over the way you smell).

Are you still on that picnic? Imagine you just took a bite out of an apple that made your tooth hurt. Take a shot of vodka, swirl it around the tooth, and voila! The toothache is gone. Some vodka pundits feel that the higher the proof, the better the results; so, if you can find 100-proof vodka, you've just found one hell of a natural remedy[86].

**Essential Oils**

There are several types of fragrances found in everyday products. Most traditional household cleaners and personal care products contain synthetic fragrances that are derived from petrochemical sources and have been synthesized to emulate natural scents. Plenty of effort goes into creating a fragrance that does not exist in nature. Food scents are common creations used to evoke a sensory response. Scents like watermelon, for instance, are not found in nature; you can taste watermelon and may pick up a hint of its scent, but the fruit itself does not create an essential oil. Nature does place limits on the type of oils we can harvest in the products we create.

Certainly not all essential oils are synthetically-derived. There are many naturally and organically-derived oils, too. The difference between natural and organic oil sources is the way they are grown. Certified organic essential oils, for instance, are grown and harvested in compliance with organic standards. A good nighttime read on this, if you would like to learn more, can be found at *www.usda.gov*.

**Walk on the Wild Side**

Try taking a walk on the wild side—you may just become mentally sharper. A recent study has found that people who take walks in natural environments see immediate improvements in their mental acuity and physical well-being; these results are in comparison to people who have no physical activity, but also to those who walk the asphalt jungles. Walking in natural environments has been shown to lower people's mental stress levels, allowing the brain a chance to recharge and improve its ability to focus on mental tasks. Walking in urban areas, on the other hand, maintains or even increases stress levels due to greater numbers of people, cars and bicyclists; in many urban settings the brain is more concerned with avoiding physical contact or injury[87].

Interacting with nature during a walk allows the brain to relax and rest. So, with a nod and thanks to Lou Reed, go and take a walk on the wild side...

## Chapter Seven: Lifestyle

"Man did not weave the web of life—he is merely a strand in it. Whatever he does to the web, he does to himself."

- Chief Seattle, *Chief of the Suquamish and Duwamish Native American Tribes*

**Human Sustainability**

Current sustainability discussions have mostly focused on energy, water, land, natural resources, and food sources. Biodiversity has been a critical part of these discussions. For the most part, the human equation in all of this has been downplayed, with the exception of how we can individually change our behaviors.[88]

Most societies skirt the subject of human sustainability (by human sustainability I'm focusing on world hunger and the exponential growth of human kind). According to the World Food Programme of the United Nations, over a billion people will go hungry because governmental aid programs aren't sustaining the hunger that exists (our own country provides fifty percent of the budget for this worldwide program). Sustainability for people is not just about feeding them, but also about providing access to clean water, health services, economic prosperity, and a biodiversity-filled environment. Human sustainability in feeding people has eroded due to economic times and human population explosions. The current population cannot be sustained with the resources we have on hand. We not only need to get our own environmental homes in order, but also to address the shear number of people on this planet. For more information on how to help, visit *www.wfp.org*.

**Airport Green Pet Peeves**

We all have our pet peeves that bother us to no end. If you Googled the term "pet peeves", you'd be amazed by the number of lists that exist: top 100, top 500, and so on. People love to make lists of things that bother them, though in all fairness, there are lists of things people love, too; but negativity seems to be what grabs people's attention the most (just watch the evening news!). So naturally, since we're talking about pet peeves, I thought I would include a few from my own list. I'll begin with air travel, as I seem to be doing a lot of it these days:

- **Water bottles that cost $4.00**. Why can't everyone be issued a stainless steel canteen instead? Then provide drinking fountains with purified water systems for people to fill them up.

- **The incessant use of disposable gloves**. TSA employees wear them constantly while checking your luggage. Since when did our luggage become contaminated?

- **Florescent lighting, bad air quality, and threat advisories**. Why not have more natural light and open windows? How long before we end the constant announcements that we are at threat advisory level orange? We get it. Do you have to keep repeating it every five minutes?

- **Non-biodegradable plastic cutlery and airports with no recycling bins**. Why not have recycling bins? Look at all the waste that accumulates each day.

- **Air freshener systems**. Though we know that bathrooms aren't always the freshest smelling places, air fresheners that make us gag and expose us to volatile organic compounds are not the solution.

I won't begin to discuss my pet peeves about actually flying. I'll save that for another time.

## No-Kill Pet Shelters Gaining

Homeless pets are a very real problem for many large metropolitan areas. Stray dogs and cats tax cities' animal control services because they are not equipped to handle the influx of homeless animals; consequently, in most cases the animals are put down in order to reduce holding and feeding costs. When there are shelters to be had, the animals are turned over; but even local shelters can't handle the constant flood of homeless pets. There is a growing alternative, though, that has captured attention.

New no-kill shelters are being developed all over the country to link homeless animals with new pet owners. One of the leading centers in the US is PAWS Chicago. PAWS (Pets are Worth Saving) has been instrumental in bringing the City of Chicago's kill rate down by more than fifty percent. The city killed 42,651 animals in 1997; by 2008, PAWS helped bring that number down to 19,288. The organization eventually hopes to achieve a no-kill scenario in Chicago[89].

Interestingly, no-kill shelters tend to be better established in northern states vs. southern states. There is even an air taxi service that brings dogs and cats from southern shelters to northern no-kill shelters. One air taxi brought over 3,000 animals into Long Island, NY alone.

No-kill shelters make a concerted effort to find good homes for abandoned animals. Pet owners who wish to adopt need to go through a screening process prior to adoption; every animal that is brought into a no-kill shelter is neutered. If you are interested in learning more about PAWS, or to adopt one of their animals, visit *www.pawschicago.org*. Animals need a break, too.

## Write Off the Cat and Dog

There is a newly proposed federal bill that would allow pet owners to deduct ownership of cats, dogs, birds, lizards, and more from their taxes. Currently, there are approximately sixty-nine million households in the US claiming to own at least one pet. The bill would allow a tax deduction of up to $3,500 per year for expenses such as pet food and veterinary care. The goal of the bill, according to the authors, is to make pet ownership more affordable and prevent people from abandoning pets during a bad economy. It may also encourage greater pet ownership and reduce the number of stray animals.

The proposed bill is known as the HAPPY Act, an acronym for Humanity and Pets Partnered Through the Years. Pet Humane Societies across the US have voiced support for the bill, as have veterinarians and pet manufacturers. Critics, though, are attacking the bill as going too far for animal lovers and taking away the focus on supporting people in need. They argue that pets are a luxury item. To house, feed, and care for an animal is indeed an expensive proposition.

It's true that pets can provide solace and mental well-being; yet relationships with people can, too (though an animal rarely talks back!). Animal owners seem to have done just fine without a tax break all these years. Taxes are now at an all time high; not just federal, but state and local taxes continue to rise despite a bad economy. Money should be used towards helping people in need—not the pet owners who have already made the choice to own a pet[90].

## Locals Control Sprawl

If you have had the opportunity to drive from South Florida into Key Largo, you have surely entered through the Jewfish Creek Bridge. The old bridge was brought down and a new one was recently built to replace it.

The locals in Key Largo, and in all of the Keys for that matter, have for many years guarded their way of life from outsiders. Not ones to welcome new influxes of immigrants and developers, they have passionately guarded their Keys; so, when a new bridge was proposed, one that would be safer and higher, the measure was brought to a local vote. The state wanted a four-lane divided highway to increase accessibility for tourists and raise their tax coffers; the locals wanted only a two-lane divided highway for increased automotive safety. Seeing that taxes would have to go up, the state allowed a vote; the locals responded with a 60-40 vote in favor of two lanes. Though the locals won, the state argued that there was a need for four lanes in order to evacuate the Keys during a Hurricane. The locals eventually agreed to two lanes going out of the Keys, but only one lane going in. They also wanted a tollbooth to slow down traffic into the Keys, creating a slight deterrent to incomers. The state said no to the tollbooth, but gave in to two lanes going out and one lane going in. These are the kinds of local victories we need to see more of[91].

## Condoms for Biological Diversity

Did you know that the US could add another hundred million people to the population by the year 2050?  We're currently at over 350 million people—can you imagine another hundred million?  How many more people can we cram into place likes California and Rhode Island?  California is already at fifty million.  Could that mean that the state would pick up another ten to twenty million people?  The State of California is the same size as Japan.  Japan already has a population of 120 million, so I guess California's population could withstand the added numbers; but would you really want to live there?  Current estimates indicate that the increase in population will likely stem from an influx of new immigrants and higher birth rates from lower income demographics.  This overpopulation will impact the world's biodiversity in plant and animal species, as well as our ability to feed people around the world.

The Center For Biological Diversity in Tucson, AZ, came up with a novel approach to attracting attention to this unchecked human propagation: on Valentine's Day, 2010, the center handed out over a hundred million condoms around the country to highlight the threat to our biodiversity.  The condoms were imprinted with pictures and descriptions of endangered species in the US.  Animals like the spotted owl and polar bear were two of the endangered species featured on the condom.  Slogans on the packages included, "Wrap with care, save the Polar Bear" and "Wear a condom now, save the Spotted Owl."  The condoms were then handed out at bars, supermarkets, schools, and public events.  I'm just waiting to see if the World Wildlife Federation gets behind it.  What a great way to raise awareness for endangered species, not to mention that these condoms make great balloon figures, too![92]

## Homeless Outreach

Santa Cruz, California, is a hot bed for political and environmental activity. For a town with 50,000 some-odd residents, it boasts over four large health food chains and several green stores that meet all of your green decorating needs. The city also has an impressive number of independent bookstores that cover a wide range of literary interests. Politically, it is an activist community with a strong sense of volunteerism and participation. It is also home to the University of California, Santa Cruz.

One of the most striking community features that Santa Cruz offers is its Homeless Hospitality Program. Uniformed volunteers that look like local law enforcement officers patrol the main business district offering assistance to the homeless. The volunteers try to develop an individual rapport with each person they reach out to and provide "tickets" that get them a warm meal, a shower, and a place to safely bed down for the evening. The volunteers also reach out to newcomers to help them learn the lay of the land in Santa Cruz and direct them to local community groups that can help. For a small city of 50,000, there are two Salvation Army locations and a local civic center that provides additional services to those in need. On certain days, the civic center provides medical care, bike repairs, clothing, and social services.

The city of Santa Cruz has one of the most pro-active communities in the country in assisting homeless people without question. These types of community programs need to be emulated by other cities in dealing with their homeless populations.[93]

### Finding Your Inspiration

Have you found a passion for something that provides you with a greater sense of self? Do you have people and things that inspire you? For some of us, passion and inspiration can be found in writing, painting, fishing, skydiving, or any other creative or non-creative endeavor; otherwise, you may be one of the fortunate people who have found inspiration in daily work. Sometimes you may not even know you have a passion for something because you just keep doing it everyday.

I know I am digressing from my usual banter about all things environmental, but I believe that finding inspiration is also an important aspect of our existence—it can play a vital role in our mental and physical health. Health is an environmental concern and affects how we interact within our natural environment.

Recently, I had the opportunity to spend time at a workshop with Frans Lanting, an exceptional artist whose photographic images have graced the cover of numerous magazines. He is a naturalist who is dedicated to preserving and protecting the earth; his images have inspired others to protect and respect the natural world. Frans is one of the fortunate ones who get to live his dream on a daily basis. Taking part in a workshop with people who have found their inspiration can be a profound and fulfilling experience no matter where your passions lie. I highly recommend doing a workshop in support of your passions. They allow you to create friendships with like-minded people and fuel your passions through the examples of others. Find your inspiration in life and you can find a fulfillment that transcends the everyday.

**Socially Responsible Dumpster Diving**

Have you heard of the term freegans? Well, they may just be the latest craze in social responsibility. The name freegan is derived from a combination of "Free" and "Vegan". Freegans practice the art of dumpster diving (yes, dumpster diving—I did say it's a craze). They embrace practices like dumpster diving as a way of protesting unethical corporations and an increasingly wasteful consumerist society. According to *Freegan.info*, "Freegans are people who employ alternative strategies for living based on limited participation in the conventional economy and minimal consumption of resources." Dumpster diving is no longer just the domain for the homeless; even professorial Illuminati have jumped on the freegan bandwagon.

University of California, San Diego biology professor, Dr. Milton Saier, a self-proclaimed freegan of thirty years, takes his students on dumpster diving outings. He shares his insights on freeganism and provides his students with dumpster diving lessons. I've previously written about the notion of freecycling, a phenomenon that enables people to give unneeded but usable items to others at no cost. Whether you call it ethical shopping, sustainable giving or just plain giving stuff away for free, freecycling is also a growing movement in the US.

The amazing thing about dumpster diving is that there are *YouTube* videos out there on how to do it (I'm a little surprised that a vegan would dumpster dive for wilted lettuce!); but the notion of reclaiming discarded consumer items has plenty of merit. Dumpster diving isn't necessarily free, though; there are communities out there that have laws against dumpster diving. So if being a Freegan is your thing or you're planning on giving it a whirl, make sure you're doing it legally; no need to get thrown into the can for diving into a dumpster.

**European Sensibilities**

Smart cars, hotel key cards that turn the lights on and off, long dinners, credit card toll roads, on-time trains, and hassle-free airport security are all things that make me take notice of Europe's approach to living.

We here in the US live in a homogenous society: similar subdivisions, strip malls, and restaurants chains provide us with a sense of familiarity and security; Europe, on the other hand, is not homogenous.  They celebrate their differences (most of the time); instead, we ridicule our respective regional differences (a bold statement I know, but just watch comedians on HBO).  I was recently traveling from Germany to Greece, on to France and finally to England.  What I found to be the common thread linking the four countries is their approaches to living.  People in these countries take a long time to eat and do it much later then we do.  Fresh ingredients; no rushing.  It's the slowness in their eating and their approach to food preparation that makes you notice the differences between their countries and ours.  Eating slowly is said to be healthier for you.  But, it also demands greater interaction with your hosts, and family—all good things in my opinion.  The Europeans also drive smaller vehicles.  Even their delivery trucks are smaller.  Their highways have less traffic because their tolls are very high, and you can even pay for passage with your credit card (something we should consider doing here in the States).

The Europeans have a very different approach to living then we do, and it's not such a bad thing to take on some of their sensibilities.

## How Do We Feed Nine Million People?

Current population trends point to a world inhabited by nine billion people by the year 2050. That means that by 2050 the planet would need to sustain an additional two billion people (right now there are already one billion people who go hungry each day).

A recent article written by nine scientists in *Science* magazine claim that we can indeed feed nine billion people if we make some major changes in how we currently feed the world. Meat, primarily hamburger, is consumed at much higher rates in the west than in eastern and developing countries. We cannot accommodate a western-style meat diet to feed that many people. This doesn't necessarily require that we move to an all-vegetarian diet, but rather one that limits meat to three times a week at most. If the limits weren't voluntarily adopted by the west, we would need to find a landmass the size of Brazil to grow cattle to feed everyone. Fisheries are also under tremendous strain with the current population. There would be a greater need to build large fish farms to handle the increased demand.

Scientists further argue that we would need to allow genetically modified organisms to become more mainstream. Higher crop yields are the only possibility in being able to feed the growing population; that means GMO seeds that are controlled by large corporations. The article states that the world's climate changes must cease in order to accommodate the additional people. Unfortunately, more people means more methane and fossil fuel consumption; biodiversity would be at risk and global temperatures would have nowhere to go but up. To reduce our numbers, we need to focus on family planning and educating developing countries, and our own country, to limit births and move towards zero population growth. Such an increase in population and GMOs invading all food types are not attractive prospects for humanity; they will only compound the problems that already exist in the environment[94].

## May Day

May Day has a long storied history. It currently celebrates the social and economic achievements of labor movements around the world, particularly in socialist and communist states, and is synonymous with International Worker's Day[95] (not something you see heavily covered in your local evening news.) US labor organizations also support the celebration of May Day. In pagan times, particularly in the northern parts of Europe, it became a day to celebrate the halfway point between the spring equinox and the summer solstice. As Christianity sought to assimilate pagan holidays unto its own, May Day became a day to celebrate the Virgin Mary. The Maypole was also an ancient tradition used to celebrate the first days of spring and of planting. May baskets used to be given on the first day of the month. These baskets would contain nuts, candies, and small items to celebrate the first day of May and the coming days of summer[96]. Whichever celebration floats your boat, we wish you a happy May Day. Long live the Revolution!

**Memorial Day**

Whether you're a Hawk or Dove, our Memorial Day holidays must always remember those who gave their lives for this country. It doesn't matter whether or not you agree with the current wars we are involved in; we must remember that men and women haven given their lives for us, lives that have been given so that this country and the ideals we've believed in can continue to endure. Despite the loss of liberties through the Patriot Act, we must continue to hold dear the freedoms that have been fought for all through the years.

Every war is different and every opinion about our involvement matters; but when we lose a son or daughter so that our rights can continue, we must honor them always. I'm not sure we should do it in a Memorial parade with fire trucks, bands and floats, but perhaps in a more compelling way. We can instead visit their graves and the memorials built to honor their legacies. A day off to remember and reflect should be in order to pay our respects to them and to their families—not just another day off from work at the neighborhood bbq.

**Green Thanksgiving**

No, this is not going to be a soliloquy on being a vegetarian for Thanksgiving or giving a reprieve to what once could have been our national bird, the turkey. The topic of greening Thanksgiving covers a wider range of lifestyle issues that we can all be mindful of.

With a nod to vegans and vegetarians, not eating the turkey is one of the greenest steps you can take; but for many out there, Thanksgiving without the bird just doesn't seem right.  So, if you're going to have turkey, look for free range, organically-fed birds.  Buying a local bird is even better (they're not always easy to find, but they do exist).  It's a great way to cut down on your carbon footprint for the holiday.

Travel is another big deal during Thanksgiving.  Just getting to and from holiday events can add to your carbon footprint (not to mention your stress level!). Instead, consider centralizing family gatherings to cut down on travel.

As you know, Thanksgiving is a big holiday for the NFL.  If you plan on watching the game, trying watching it on just one TV set in the house, not on all of them; better yet, turn it off and go have your own Turkey Bowl game outside.  There are also numerous parades to go to on Thanksgiving.  Having a family get together at an outdoor event not only brings everyone together, it connects you to your own community.

Now go enjoy the holiday—and try not to eat too much.

**Summer Travel Plans**

Most of us think about summer with great anticipation: beautiful weather, long days, and a sense of adventure beckon us to hit the open road. We look forward to a summer vacation that takes us away from the everyday; and summer vacation can actually be one of the more inexpensive trips you take, depending on just how far you want to drive or fly.

The National Wildlife Refuge System is the world's premier system of public lands and waters set aside to conserve America's fish, wildlife, and plants. The US Fish and Wildlife Service manage it, and the system has multiple refuge parks in every state in the union. President Theodore Roosevelt started the system in 1903 when he designated Florida's Pelican Island as the first wildlife refuge. The service manages 150 million acres, 551 national wildlife refuges, and thirty-seven wetland management districts. No matter what part of the country you live in, you can easily drive to a wildlife refuge in your state; some refuges can be as close as thirty minutes from your home, while others can be as far away as a five-hour drive. The US Wildlife Service entrance fee is a bargain; for twenty-five dollars you buy their annual Duck Stamp, which provides you with access to all Wildlife Refuges in the system. The stamp can be purchased from bass shops, the US Postal Service, or from the US Fish and Wildlife Service. A complete list of locations and information on the National Wildlife Refuge System can be found at *http://www.fws.gov/refuges/index.html*

**Summer Green**

Heading to the beach or local pool next summer? If the answer to this question is yes, try to see how green you can be while you're there. I know that most of us just want to get to the water and dive right in (unless your first love is getting that golden summer tan); but we need to be green about it.

First, if you're headed to the beach, make sure the beaches are open and healthy. The EPA has a website that let's you know about every shoreline's health status. Make sure to check with the local lifeguards, too. Beyond the water's bacteria count, you need to take care of your skin. You can start by looking for paraben-free sunscreen products. There are an increasing number of paraben-free sunscreens sold at specialty stores, co-ops, drugstores, and even large grocery stores. Are you in search of beachwear? Look for long lasting garments; better yet, try skinny-dipping (not that many of us would look very good doing that).

Every beach or pool outing calls for a good cooler. Just make sure the cooler isn't styrofoam. It's worth investing in a long lasting quality-built cooler. While you're at it, why not stock up on local beer and organically-grown wine? There are natural sodas to buy, too. Water is always a necessity, but try to lose the plastic bottle variety, especially if you've got good tap water.

Now, go enjoy the summer and do your part to make it a greener one. See you at the shore!

## Green Age Differences

IBM recently conducted a survey to discover which generation amongst us would be considered the greenest. Right out of the gate most of us, including myself, would have looked to the youngest generation out there, Generation Y. Despite the fact that they may be the most environmentally aware, they are also the least green in their daily habits. They have the highest awareness, but are the biggest wasters of energy and water in the country. IBM's survey says that Generation Y is the most concerned generation out there. Yet, somehow there is a disconnect between awareness and action.

The IBM survey had a few interesting highlights on their Generation Y findings. One such highlight was the fact that fifty-five percent of them correctly guessed on which used more energy, the electric dryer or incandescent lightbulb. They also admitted that they run the water prior to entering the shower (who doesn't?) and tend to leave the water running as they brush their teeth. Then again, the final explanation could be that they've left home and we're not there to remind them to turn the water off[97].

**Plastic vs. Glass**

I recently did a little digging on the merits of glass versus plastic. As some of you probably know, plastic bottles contain Bisphenol A, an ingredient linked to cancers like prostate and ovarian cancer. The ingredient has also been shown to cause an early onset of puberty in children. Many plastic containers, primarily plastic numbers 3-5, are not recyclable. White-colored number 2 plastics tend not to be recycled either since they are color tinted.

The manufacturing of plastic bottles consumes approximately four percent of our country's annual oil use. Plastic is lighter and costs less to transport over heavier glass bottles, and they rarely break. Plastic bottles are manufactured through molding machines that run on electricity, whereas glass bottles are made in furnaces that run on natural gas. As opposed to the often one-time use of plastic, glass bottles are recycled back into commerce on an average of eight times prior to disposal. A recent study conducted in France found that glass bottles that are recycled back into commerce employ more people than the plastics industry.[98]

We also know that it takes a longtime for plastic to breakdown—something to the tune of a thousand years; on the other hand, glass biodegrades much more quickly. Yes, plastic is more convenient for those on the go (imagine taking a glass water bottle on a bike trip!); but in my opinion, when you drink something out of a glass, it just tastes better. Both types of bottles have their place, but with all things considered, glass just makes more environmental sense.

**America the Beautiful**

Once you get past the urban sprawl, strip malls, and tract housing where the majority of us live, venture forth and see America the Beautiful. Don't just see it on a *National Geographic* special or at an IMAX movie—get out and see it in person; drive it, don't fly. Re-read Jack Kerouac and get inspired. See the expansiveness of our country.

This week I'm in Monument Valley, which straddles the Utah-Arizona borders. The Valley is on the Navajo Nation lands and has long been protected by them. It's not part of our national park service; it's part of theirs. The Valley is really quite a site; the expanse and views are breathtaking. The tranquility of the land is truly awe-inspiring.

Monument Valley is just one place to visit in America. Now, with Earth Day approaching, I'd recommend putting together your bucket list of places to visit. Take some time and relearn your American geography. Its good for the economy but, more importantly, it's good for you and your family.

## Chapter Eight: Consumer

"Our personal consumer choices have ecological, social and spiritual consequences. It is time to re-examine some of our deeply held notions that underlie our lifestyles."

- David Suzuki, *scientist, environmentalist and broadcaster*

## Kitchen Scouring

Kitchen scouring is not what we would call a hot topic these days; most of us use a dishwasher or go about cleaning dishes by hand. If you tend to do your pots and pans by hand, then you most likely use a scourer to remove food and grime. Kitchen scourers come in all shapes and sizes. The materials used in scourers usually range from synthetic to bamboo. Steel wool is also a common household item, but once you've used it, your hands can get fairly irritated from the harsh chemicals it contains. Many of the new sustainable scourers are made from luffas, a vegetable that can be found in the Philippines. As a vegetable, they can be stir fried or used in stews. Farmers who grow luffas allow them to go to seed for the following season's crop; those that are harvested for scouring are allowed to grow to hand size. They become supple in water and can easily be bent to reach into corners when cleaning. Luffas and other natural scourers will completely biodegrade once tossed into a compost bin or landfill. Unlike synthetic scourers, plant-based scourers will not contribute to global warming or deplete finite resources. When done right, kitchen scouring actually helps reduce $CO_2$ output. Who would have thought?

**Greening your Brew**

The Fourth of July almost always guarantees that America will fire up BBQs and celebrate the holiday in ways that only we can appreciate: lots of brats, dogs, burgers, and brew. With upcoming celebrations in mind, let's try to "green" some of these traditions. Vegans will not be happy about the meat, so let's do something about the brew instead.

Before heading out to the local libations store, do a little research and support your local brewery. Look for breweries in your area that support fair trade and perhaps have a brew that uses organically-certified ingredients. By supporting a local company, you help cut down on the distance the brew traveled to get to your backyard. Supporting fair trade ingredients guarantees a fair wage and a good work environment for foreign workers. Organic ingredients help lower our dependency on man-made fertilizers and pesticides and also help local farmers, who seem to be doing the greater share of organic farming.

Support local companies and be responsible; make sure you've got a designated driver lined up. Have a great holiday, and Happy Birthday America!

**Green Lawns are a Push Away**

Spring has sprung around the country; well, at least in the Midwestern and Northeastern parts of the US. With spring's arrival, one's fancy often turns to lawn maintenance (that's probably not what you thought I was going to say). Reviving one's lawn may not be such a hot topic at your neighbor's cocktail party, but it is something all homeowners grapple with at winter's end. In the spring, why not make a great green choice in lawn maintenance that will reduce your carbon footprint and provide you with some real exercise benefits, too: buy a push mower. I know they are not the most glamorous, but they certainly make an environmental difference when it comes to lawn maintenance. Manufacturers like Brill, Scotts, and Mascot make a range of affordable push mowers; plus, for all you guys out there that like add-ons, they provide them as well.

If push mowers are not for you, try a solar-powered mower from Solaris; from what I hear, the Solaris model is pretty quiet. If you just can't seem to do the push mower or solar mower, then try an electric-powered one. One downside to all-electric mowers, though, is the cord that accompanies them out into the yard; but whichever mower you choose, you are making a significant contribution to lowering your carbon footprint, and get the added reduction in noise, too. Your neighbors will love you for the noise reduction, especially if you decide to cut the grass on Sunday morning.

**Greening Mother's Day**

As you may know, Mother's Day has a long and storied history dating all the way back to ancient Greece. The holiday actually emerged from a custom of mother-worship, which was part of a festival honoring Cybele, a great mother of Greek gods. The city of Rome picked up on the idea and began celebrating its own Mother's Day holiday around the Vernal Equinox. In almost every part of the world you will find a history to celebrate a day honoring mothers. In 1870, the "Mother's Day Proclamation", written by Julia Ward Howe, was one of the earliest calls to a Mother's Day celebration here in the US.

In anticipation of Mother's Day, I thought I would provide a few eco gift ideas for all of us guys who can at times be pretty clueless about gift giving. These gift ideas are a nod to being less oil dependent. Let's give our mothers gifts that are not derived from oil sources. We can start with fair trade 100% organic chocolates from Fine and Raw Chocolate. These chocolates are devoid of sugar, dairy, and unnatural additives, and can be found on the company's website at *fineandraw.com*. Jewelry is always a good bet, too. This year I'd go for glass bead or acacia berry necklaces created from sustainable resources. A good shopping source for jewelry is *inhabitatshop.com*. Though kitchen items won't win you many points, I can't help recommending some alternatives to plastic bowls, cups, and dishes. Try getting your household to switch to bowls made of materials like Pyrex, bamboo, ceramic or cork; and get rid of plastic, which will help Mother Earth and your home.

Obviously, flowers are a big deal on Mother's Day. Look to buy flowers that are grown locally (it doesn't always have to be roses); and if you fail in doing anything for Mother's Day, then make sure you've made reservations for brunch or dinner—and keep it local!

**Green Gadget Conference Wows**

Have you heard of conferences such as Comic Con, Game Con, Food Con, and Botox Con (just kidding on that one!)? New York City hosted a Green Gadgets Con in which designers showcased their ideas and prototypes for green-friendly gadgets as part of a green design competition. My favorite designs included a solar powered tent, a battery charging station, a battery-free computer mouse, a power-generated rocking chair, and a smart phone application that tells the distance your store-bought produce traveled to get to you.

As part of its sponsorship of the Glastonbury performing-arts festival, British telecommunications giant, Orange, commissioned the US design firm, Kaleidoscope, to create a concept for a solar tent. They wanted to create a product that would make a positive environmental impact while giving festival-goers a place to hang out and keep their gadgets fully charged. Designer Adele Peters came up with a computer mouse made of cork that generates its own power simply by moving around a desktop. The mouse is all hand-powered and doesn't require batteries. Meanwhile, Pensa, a New York design firm, created a device called the InCharge battery station. The InCharge battery station encourages the use of rechargeable batteries, with a self-sorting design that sends power only to batteries that need it. The sorter can handle AAA, AA, C, and D batteries.

One of the coolest summer furniture ideas I've come across is a rocking chair that generates power through its own rocking motion. It also provides trickle charging for USB devices, which can be powered through a trickle charger.

The smart phone application took first place at the Green Gadget Conference. Using a smart phone camera, you can take a picture of the bar code on your food item; using a GPS to locate your position, the bar code is scanned and then the application tells you just how far your food traveled to get to you.  One of the advantages of the application is that it allows you to make informed and sustainable buying decisions.

It's too bad such conferences can't take their showcased products on the road.  It would be great if we could expose the public to such novel ideas and build public awareness and demand to support these efforts.

## Consumers in a Bind Over Fish

Our annual spring break family trip has finally brought us to San Francisco. While in the city, a news article regarding the safety of the seafood in the Bay Area caught my attention (fortunately, this was well after a cable car ride to Fisherman's Wharf and a stroll through its various fish market stands). According to the article, an undercover investigative team discovered that of twenty-four individual pieces of fish bought around town, ten of them had mercury contamination higher than legal standards. The news article did acknowledge, though, that the team's investigation sought out fish known to have high mercury content, such as tuna and swordfish. Despite the news revelation, it still raises the worrisome question as to whether the fish we buy locally is also contaminated at above legal levels.

The San Francisco Bay is not alone in offering contaminated fish for consumption. In fact, mercury levels found in the Bay Area are lower than those found in the Great Lakes. Unfortunately, the mercury levels found in the fish brought into San Francisco contained enough contaminants to fulfill a person's exposure to mercury for a whole year. Retailers who sell contaminated fish now find themselves in a bind: they themselves do not go out and harvest the fish, but instead depend on suppliers to bring in their catch. In turn, as consumers, we are dependent on an over-taxed regulatory system to monitor these harvests. Fish are an important food group for people. Our polluted waterways impact the fish we eat, and relying on farm-raised fish brings on a whole other health issue surrounding the antibiotics used on those fish farms. We need to escalate our efforts at protecting our waterways. Otherwise, news articles like the one in San Francisco will become the norm[99].

## Personal Changes Can Cut Emissions

Did you know that inexpensive personal actions can reduce total US carbon emissions by fifteen percent, according to a recent study? The new study was presented by the National Resources Defense Council and the Garrison Institute's Climate Mind Behavior Project at a symposium on "Climate, Mind, and Behavior" in New York City. The study's analysis reveals that Americans can reduce their carbon pollution by fifteen percent—roughly one billion tons of global warming pollution—through personal actions, which require little to no cost to consumers.

Most of the focus by government and the media has been on large-scale projects to reduce the US contribution in carbon emission reductions. Even as the country struggles to meet lower carbon emissions, personal contributions by consumers can have an immediate impact just by making a few changes in our daily lives. The analysis suggests certain behavioral changes people can make to reach these goals; quite frankly, the suggested behavior modifications are effortless and painless for consumers. Some of the steps suggested include reducing unwanted catalog subscriptions, decreasing vehicle idling, using a programmable thermostat, replacing seven light bulbs with CFLs, setting computers to hibernate mode, shutting off unused lights, and eating poultry in place of red meat two days per week.

All of the recommendations offered in the study are available to be adopted immediately at little or no cost to the consumer and will reduce not only emissions, but home energy, transportation, and food costs, too. If we all took to heart the above-mentioned steps, imagine what we could do to lower our carbon emissions. If we lowered our emissions by fifteen percent, it would be like eliminating all of the global emissions of the UK and Saudi Arabia combined. Think about it[100].

## Least Green Consumers

NatGeo released their annual report for 2009, which ranked the countries with the greenest consumers. Where might you think we came in as a country? We were dead last. The US was last primarily in the extent to which we build environmentally sound homes and in our use of public transportation.

In all fairness to US, we do not have the sheer number of public transportation systems that Europe and Japan have in place, so being last in using public transportation use seems a bit like an oxymoron in my opinion. How can you use it if it doesn't exist? We have bus lines, but not in every city; and cities that have great public transit systems such as New York, Chicago, Boston, and Philadelphia, are running at almost full capacity.

In my opinion, if you build more public transportation, people will use it—housing is another issue. Many homebuilders have declared bankruptcy in recent years. On the other hand, new homes being built are seeing a rise in the use of green building materials and more people are trying to get their homes off the grid; but they are the exception, not the rule. We've got a lot of work to do in our own backyard. It might as well start with you[101].

## Climate Changes Impacting Coffee

We all love a cup of coffee in the morning; well, many of us do at least. What would we do, though, if that cup of java became prohibitively more expensive, (even more so than it already is!), or unavailable due to worldwide climate change? It seems that the country of Kenya is already grappling with that question. Climate change has affected Kenyan coffee production through unpredictable rainfall patterns and excessive droughts, making crop management and disease control a nightmare; intermittent rainfall has caused a terrible bout of the Coffee Berry Disease and has in turn cut Kenya's output by twenty-three percent. The intermittent rainfall patterns have also contributed to extended drought periods and high temperatures.

Coffee operates within a very narrow temperature range of nineteen to twenty-five degrees Celsius. When you start getting temperatures above that range, it affects photosynthesis and can cause trees to dry up and wilt. For coffee to flower, it needs several months of dry weather followed by showers; but in 2009, Kenya had rains in January, which is normally a very dry month when bushes undergo what is known as "stress" before they begin to flower. Because of the unpredictable weather, bushes began flowering too early and having coffee berries at different stages of maturity, which meant farmers had to hire labor through most of the year to pick very few kilos of coffee. Current coffee trees cannot determine the season since each tree has beans of different ages due to the fluctuations in climate. All of these factors have reduced harvests and raised costs. So if you think a four-dollar latte is expensive, what will you think of climate change if that cup of java hits ten dollars[102]?

## New Watered Down Standards

The Natural Products Association, in conjunction with Whole Foods and Clorox, has developed a minimum standard to allow companies to claim that their products are natural. Not surprisingly, Clorox's Green Works products are being touted as completely natural; S.C. Johnson claims the same on their green product line.

Whole Foods, while trying to do right by their consumers, has clamored for some kind of standard to help them communicate to their consumers that the products they sell are natural. Their intentions are well meaning, but the reality of this new standard falls short for consumers and actually misrepresents what can be construed as a natural product. According to the Natural Products Association, it's acceptable if only sixty percent of a product line meets these new standards. It's also acceptable if only ninety-five percent of the ingredients used in your product are considered natural (water is not counted in these standards); so it's OK to use petroleum-based ingredients if they are biodegradable and if there are no alternatives.

The funny thing is that the NPA does not look to see if there are substitutions for the ingredients these manufacturers claim they cannot source from natural suppliers; they instead expect the industry and manufacturers to self-regulate. Companies like Clorox, Seventh Generation, Method, and Mrs. Meyer's like to claim that they are all-natural brands, but if you actually look at the ingredient lists and the dyes and perfumes included, you can see that they are not selling you one hundred percent all-natural products. Getting the NPA to endorse these products only adds a false legitimacy and continues the green washing that exists in the marketplace[103].

## Local Foods

The herald of summer for me has always been the first local farmer's market. Living in a cooler climate makes it difficult to support locally grown food; but with summer's arrival, it changes everything.

Adding locally-produced food to your grocery list assists the environment in many ways, besides just being healthier for you; it lowers transportation costs, supports local family farmers, and encourages community interaction. Finding a local farmer's market is fairly simple. Just look through local newspaper ads, seasonal signage, or through word of mouth. A faster way would be to do an online search for the farmer's market closest to you.

June is one of my favorite months for the local farmers market, particularly in New Jersey, where fresh blueberries at this time of the year are just amazing. If for some reason you can't get to a local farmers market but you have some free yard space, grow your own. There is nothing like a garden fresh tomato—assuming of course that you like tomatoes.

## New Electric Car Options

US consumers will soon have true electric car options. With thoughts of the devastating oil spill in the Gulf reminding us of the volatile nature of our current oil-dependency, having true electric car choices is a welcome reprieve.

Nissan's Leaf and Chevy's Volt will both be available to consumers. The Leaf is a one hundred percent all-electric car, while the Volt runs on an electric motor but has a gas engine to help extend its range and recharge its battery when needed. Both vehicles are sedan-style, four-door offerings[104]. Another all-electric option is coming from the car company Think. Think cars is a Norwegian automaker that has strong ties to the US; it received $118 million in US energy stimulus grants to open a plant in Elkhart, Indiana. The new plant will have the capacity to produce 60,000 Think cars each year. Initial Think cars will soon hit New York City, with additional rollouts planned for Chicago, Los Angeles, San Francisco, and San Diego. Think City, the company's moniker for the car, is highway-legal and runs exclusively on a rechargeable lithium ion battery-powered train. It has a range of about 112 miles on a single charge and a top speed of sixty miles per hour. While it can take up to eight hours to fully charge from a 110-volt household outlet, the Think City can charge to eighty percent capacity in fifteen minutes at a fast-charging 220-volt station. Think recently picked up a contract for a Brazilian utility company to supply them with a fleet of Think City cars[105].

**Planned Obsolescence**

Reuse, recycle, restore has long been the environmentalist mantra; finding ways to extend the lifecycle of consumables just makes green sense. This brings me to the topic of planned obsolescence. Planned obsolescence is a term used in industrial design that refers to the deliberate designing of a product so that it has a limited lifespan and becomes non-functional after a certain period of time. My question is what kinds of consumables can we use to eliminate planned obsolescence? Are there practical designs of today's consumables that can be altered to avoid obsolescence on these goods? In trying to eradicate obsolescence, we extend the finite resources we have on hand. Think building for longevity—saving these resources for future generations. Can you imagine a car or plane that could last multiple lifetimes? What about houses that can stand for generations instead of four-sided, cheaply-built cookie cutter homes; and highways that last more than fifty years? What about appliances that work for many years rather than ones that need regular replacing? As a consumer, would you pay more for something that would last a longtime; or have we become such finicky creatures that we regularly need to change what we have for our own satisfaction? No doubt the media and advertisers play a role in this (though that's a topic for another time). We need to re-examine our priorities if we are going to be able to provide for future generations.

**Customer Service**

Has customer service become a thing of the past? Has corporate America, or global conglomerates for that matter, given up on customer service? Has the term "customer service" become an oxymoron? For those of us who have had the pleasure of being rerouted to a foreign country to speak with a customer service representative who doesn't really speak English, you know what I'm talking about.

I just spent forty-five minutes trying to elicit a product return authorization number from a customer service representative at a large technology company—incredibly frustrating! I was placed on hold every three minutes so that he could check their " resources" to see if they could do anything to help me. It then took fifteen minutes to get a supervisor to approve the return and another ten minutes of being told from their hardware department what to expect during the return process. Why do companies do this? Is it done to frustrate customers to the point of hanging up? Is it a conspiracy?

I have always used Apple hardware—and now I know why. Just walk in to an Apple store and they'll take care of any problem you have. Some companies have not lost sight of customer service; others, apparently, just want you to learn a new foreign language.

## Wal-Mart's Sustainability

The second largest corporation in the world has given its suppliers marching orders in tune with its new sustainability programs. Wal-Mart/Sam's Clubs have insisted on lowering the amount of packaging used in delivering products, increasing the value benefits to the consumer.

The large retail chains have developed a vendor scorecard to detail how the vendor's products are created, packed, and shipped. How your scorecard looks to the retail chains has an impact on the business you do with them. Wal-Mart is concerned with the carbon footprint of each item that comes into its stores; where the product ships from and how it is shipped are important considerations for them. New stores are being built with green technologies in heating, cooling, and lighting, and they have reduced the amount of paper being used internally, even to the point of reducing the number of business cards issued to employees.

Like all companies, there are always areas to improve. Finding ways to reduce their CO2 emissions and get their stores off the grid would go a long way in setting a corporate example for other Fortune 500 companies. When Wal-Mart talks, their vendors and the communities they are involved with, listen[106].

## Retail Pricing

Why do retail prices vary from store chain to store chain? These are comments that we receive from consumers and can easily explain. The majority of consumables move through a variety of distribution channels before arriving in your home. Club stores like Sam's Club and COSTCO buy in large volumes directly from a manufacturer and have lower mark ups and larger sizes to help lower the cost to you.

Regular grocery stores tend to buy in two ways: direct from the manufacturer or through a distributor (a distributor will buy from the manufacturer and re-sell it to a grocery store). The second process leads to higher retail prices on the grocer's shelves. Most drugstores buy directly from the manufacturer, for instance, whereas natural health food stores tend to buy through distributors. This leads to different retail pricing. Markups vary from distributor to retailer, and each retailer sets their own markups, which are not uniform. Not surprisingly, it can be quite frustrating and confusing to a consumer. You can imagine what it does for companies in managing this process. We're just as frustrated as the next guy!

## Coffee Cup Sustainability

We all love a cup of coffee or tea in the morning; at least everyone I know does. For some, that could mean our own home brew or a quick stop to our favorite local coffee shop. Personally, one of my favorite morning coffee places is Caribou Coffee. Caribou Coffee is small chain out of Minneapolis with nowhere near the coverage that Starbucks provides. As a chain, the company is trying to do their part in getting people to reduce the amount of paper they use for their morning brew. One of the company's ideas has been to try weaning customers off of corrugated coffee cupholders. Caribou has come up with a permanent fix for cupholders, if you're willing to spend a few extra bucks. They are selling reusable coffee cupholders in a variety of colors and styles. The company estimates that over 1.1 billion disposable coffee cupholders are used annually in the United States. I'm not sure what percentage of them come from one hundred percent recycled paper, but they certainly add to the number of trees being cut down and the amount of waste being added to landfills each year[107].

The only downside to buying a reusable holder is that you have to remember to bring it with you the next time you buy coffee; and if you lose it, it's a costly process to keep replacing it. It does make for a great office gift or corporate giveaway, though. I just like the fact that a chain of coffeehouses is trying to lessen their waste stream, even though it comes out of our pockets to achieve it.[108]

## Corn Demand Exceeding Supply

According to the US Agriculture Department, American farmers can no longer meet the growing demand for corn; exporters, livestock feeders, and ethanol makers are going through the US corn stockpile faster than farmers can grow the crops. Despite record crops in two of the past three years, the US Department of Agriculture estimates that corn carryover will shrink to the lowest level since 2006/07. In a monthly look at crop supply and usage, the USDA estimated that 1.478 billion bushels of corn would be in US bins by August of 2010, when the marketing year ends, and 1.373 billion bushels would be on hand at the end of 2010/11. The report was anticlimactic after the USDA recently shocked markets with data showing corn plantings were smaller than expected for the year and corn consumption was far higher than expected. The data had spurred US corn futures up sixteen percent to a two-month high; but prices were still far below 2010's peak above $4.25 a bushel seen at the beginning of the year.

Our growing demand for corn can't be sustained with so many competing interests seeking it. Ethanol, which was once perceived as a savior for our mobile energy demands, can no longer be counted on to fill our vehicle demands. The need to feed the world takes precedence over filling our fuel tanks. American farmers thus far have been highly successful in providing us with sustenance; but the rising demands that competing interests are making on them is taxing their ability to satisfy everyone[109].

**Can the Brain be Green?**

Are humans naturally wired to be green-minded? Do we really care about the environment, or are we just interested in instant gratification?

The New York Times Magazine recently ran their annual Green issue, providing us with a collection of interesting hyperbole on greenness. One thing that caught my eye was the question of whether or not the brain can be green. I'm not sure Western society is geared to be green. Just think about it. How many of us played with building blocks growing up? We'd painstakingly build up the blocks and then delight in knocking them down. We love to build things, consume things, and then blow them up. We're wired to survive and take care of ourselves no matter what.

Native Americans were good at taking care of their surroundings. It seemed more important to them to protect the environment by making sure not to take more than they needed; they moved around to allow nature to replenish itself. We, on the other hand, love to take and keep taking.

As we approach another Earth Day anniversary, the question is, do we have what it takes to convert our bad habits into good ones; or do we need to be pushed to extinction in order to make an evolutionary leap towards changing our ways?

## Chapter Nine: Green

"When green is all there is to be, it can make you wonder why, but why wonder. I'm green and it'll do fine, it's beautiful, and I think it's what I want to be."

- Kermit the Frog, *Muppet character*

**One by One**

One by one, we can change the world.  We interact everyday with our environment, sometimes proactively and sometimes as an observer.  Either way, we are always a part of it, as it is part of us.  That notion got me thinking.  I had been kicking around some ideas for how people could help each other influence positive developments in their local environments; besides volunteering, working in a green business, or just being conscientious and good stewards of the Earth, I wanted to get more interactive.

I am a big fan of photography and an avid participant, too.  I believe that a single photograph can change the world; a single image can alter people's perceptions and galvanize them into action.  Our human history is filled with images that can capture a moment in history or tell a story of what a person is feeling.  In some way, everyone has taken a picture at some point in his or her life, whether with an instamatic, disposable, cell phone, or digital SLR camera.  With that in mind, I started a new section in the Earthy Report.  The section offers a chance for avid amateurs, seasoned professionals, students, and anyone who takes a camera out at parties and events, to show off their pictures.  I do have one catch, though: the images need to be related to the environment and have a short description attached.  The environmental image can be anything that is important to that person: their bedroom, their backyard, the nearest landfill, or a breathtaking image of a person or place they love.  They could be working on a project to protect a local ecosystem or trying to get a polluter to change their ways.  It doesn't matter; all I'm looking for is a chance to share and encourage others to be aware of and document their environment.  It's a way to capture the present and protect the future; and who knows, it could just change our world one by one.

**Renewable Biomass**

A newspaper article from Princeton University recently discussed the issues surrounding biofuels derived from biomass; the article's focus was on the balance of biofuel production, food security, and emissions reduction. The authors of the article concluded that the biofuels industry must focus on the major sources of renewable biomass (the raw materials used to generate biofuels): perennial plants grown on degraded lands abandoned from agricultural use; crop residues; sustainably-harvested wood and forest residues; double crop; mixed cropping systems; and municipal and industrial waste. The authors estimated that these sources of biomass, 500 million tons per year, could produce enough fuel to meet the majority of US transportation requirements.

The newspaper article also considered the amount of CO2 reduction and new creation during the production of these fuels. It reported that the new CO2 levels would be lower than the current amount of CO2 being produced. Another consideration discussed was the balancing act between using biomass for biofuels and the need to feed humanity on a global scale. The authors call for biofuel production to transition away from using food crops like corn and instead focus on biomass sources that could be produced with less impact on people and on the environment. The newspaper article will be published in the journal of Science; the title of the paper is, *"Beneficial Biofuels --The Food, Energy and Environment Trilemma.*[110]

## Seed Banks

London's Kew Royal Botanic Gardens recently achieved a milestone in plant conservation. Kew celebrated the acquisition of their 24,200th plant species: a pink wild forest banana from China, which is a staple for wild Asian elephants.

Kew Gardens started a project back in 2000 to collect ten percent of the world's plant species. The conservation effort has been in order to secure a genetic seed bank of the world's wild plant species and safeguard plant biodiversity from extinction. The most recent acquisition by Kew, the Yunnan banana, has increasingly come under threat in the wild due to its jungle habitation being cleared for commercial agriculture. Kew announced that by the year 2020, their goal is to acquire and safeguard twenty-five percent of the world's plant species. According to them, between 60,000 and 100,000 species of plants are threatened with extinction, roughly one quarter of the world's diverse plant life.

Kew's effort is the world's largest conservation project on record. The time and cost required to accumulate these plants is quite staggering (to acquire and house a plant costs roughly $5,000).

The organization sponsors global efforts to educate and train farmers in areas where plant species are under the greatest threat. They have targeted areas in West Africa and Asia as the two most plant-diverse regions in need of protection. Plant biodiversity is key to life on this planet and is a cornerstone of our existence. Losing it would be pure folly[111].

**How Green is Green Housing?**

What is your definition of a green home? Is it a home that is energy efficient and built from sustainable materials? Green architects and environmentalists would argue that a green home needs to meet those criteria; but I would offer one more criterion, which tends to be overlooked: the actual footprint of a new home. The desire to own something new runs deep in our society. We can't dissuade people from building new homes, but just how big should a new home be? Can we call a "McMansion" a green home, even if it has been built to green standards?

Taking up land to build new subdivisions has always been one of my pet peeves. Given the finite reality of our resources, we need to address just how big a new home should be. A 10,000 square foot home for one average family is an extreme luxury given that the average home built in the US is 2,500 square feet[112]. Our standard of an average US-built home of 2,500 square feet far exceeds the world average of 1,100 square feet for a home.

Green doesn't just mean water conserving showerheads and compostable toilets. The footprint of a home needs to be added into our definition of what constitutes a green home. Not adding the size of a new home into the green equation overlooks the fact that these big homes use more energy and require more resources to maintain them.

## Grassahol Fuel

Ethanol has been the gas alternative darling for some time now; yet recent studies have shown that ethanol requires large amounts of water and energy to create. Besides the amount of water and energy needed, it also takes away corn that could be used to feed people. There may be an alternative, though. New studies have been showing the potential of using grass for fuel; not the ordinary Kentucky Blue grass we are all familiar with, but switchback grass found in the plains.

Switchback grass is a hardy plant that grows from Texas throughout the plains states. It grows like a weed and is not a plant often utilized in human consumption. Researchers have been converting the grass into what they call "grassahol" and have found that it requires less water and energy to maintain. Also, the yield of fuel that switchback grass creates is greater than corn ethanol[113].

Grassahol has been used to fuel a variety of car engines in testing its efficacy. Estimates claim that US switchback grassahol could meet up to fifty percent of our vehicle fuel needs. The road to freeing us from imported oil might just be closer than we think.

## Is Your Lawn Bad For the Environment?

A new study from the University of California at Irvine has determined that grass lawns are polluting the environment. The study, led by Amy Townsend-Small, an Earth system science postdoctoral researcher at UC Irvine, specifically looked at the gas emissions created by lawn mowing, leaf blowing, irrigation, lawn fertilizer manufacturing, and the nitrous oxide released from soil after fertilization. The researchers found that the amount of carbon dioxide emitted from lawn-related maintenance was four times the amount of carbon naturally collected and stored by the lawn itself; the collective lawn maintenance emissions far outweighed the amount of carbon the lawns removed from the atmosphere through photosynthesis and storage of carbon in soil. As Townsend-Small said in a statement, "Lawns look great—they're nice and green and healthy, and they're photosynthesizing a lot of organic carbon. But the carbon-storing benefits of lawns are counteracted by fuel consumption." The UC Irvine researchers didn't simply calculate statistical averages to come to their conclusion. At four parks in the Irvine area, they measured soil samples for carbon sequestration—air above the lawns for nitrous oxide emissions—and fuel consumption for park maintenance; the parks included ornamental lawns, picnic areas, and athletic fields.

Advocates for creating sustainable lawns seek out native flora to determine the best approach in planting. Desert communities are reverting back to native plant species instead of trying to plant green grass that draws from scarce water sources. The trend in the past was to reduce the amount of water needed for plantings. Now the need is to discover native plant species that do not require the current regimen of lawn maintenance. I'm not sure what type of lawns we would have in the Midwest and Eastern parts of the country if Kentucky Blue Grass were eliminated. Would that mean that we would resort to native and non-native weed species? In Illinois, that could mean a return to native prairie plant species for homeowners. Who knows, after that, the Bison might make a return[114].

**Green Guilt**

In the last few years, "Green" has become the new media darling; numerous companies have benefited from being in the green media spotlight. As with all darlings, though, it seems that the luster has begun to wear off.

George Will had an interesting newspaper column that focused on an animated sitcom called *The Goode Family* and their green angst antics. "Guilty Green" was one of his highlights, and it caught my attention. Guilt is a powerful condition in human relationships. Using guilt to get you to buy green products or to be greener in your daily life is ethically disturbing. As we know, our society is built on choices. You have the right to choose how you wish to lead your life (as long as it is well within the laws we have created as a society). Using green guilt to ridicule or point fingers will not work to promote a greener world; yet, facts and scientific research will. Hopefully people will consider the proper scientific ethics when choosing green lifestyles and not be guilted into thinking that they should be—that only leads to green resentment[115].

**French Green Apples**

The French have long been accused of coming late to the green party. They have their traditional independent way of thumbing their nose at the green community. They have fought against banning fur in clothing and up until now, recycling; but on a recent trip through Paris, I sensed that the recycling attitude had changed quite a bit. Everywhere I walked there were green plastic containers that resembled green apples. At first, I thought they might have been a new design in public toilets, since "pay to pee" toilets are on every corner; but upon closer inspection, the containers had multiple outlets for depositing plastic, glass, and aluminum. The green containers are apparently all the rage around Paris. I call them the French Green Apples. Why the design of an apple, I have no idea.

You can find other cool features around the city, too. They have public areas set up so you can self-rent a bike to get around town; just drop a coin in and away you go. When you're done for the day you can drop off the bike at another bike rental station rather than having to return it to the original one. It's a great way to get around town without relying on fossil fuels. Écouter magnifique!

**Is Planting Green Being Green?**

Arbor Day is always celebrated on the last Friday in April. Perhaps by now the popularity of Earth Day has usurped it, but in my mind, Arbor Day is the original "Earth Day". When I was younger, we used to plant seedling trees in honor of the day. I don't how many trees I've planted in my life, but it has been important to me to plant them. Somehow, I thought it was good for the environment; I still do. Now, Arbor Day has become something more. It's not just about planting trees, but about caring for them and finding ways to recycle paper.

Planting a tree is a feel-good deed; it's something that will give back to us for years to come. I asked my son, who is in grade school, if his class talked about Arbor Day or planted a tree in honor of it. I got a typical blank look in return. It seems our country has some work to do on this one. So let's work towards getting back to the notion of planting trees and being a little more aware of our environment. Planting a tree is a great way to make a contribution to your local environment; it's a way of starting down the path to leading a greener lifestyle[116].

## Fall Home Efficiencies

As summer fades into a distant memory, our thoughts usually turn to bunkering down for the winter.  Fall thus becomes a great time of year to clean up your house and yard.  It's also a good time to call your local utility company and ask for a free energy home inspection (always one of our favorite things to do, I know).  Local utility companies are happy to come out and check your home's energy footprint.  They use specialized equipment to detect air leakages from windows, doors, and roofs and also examine your current heating system's efficiency, all of which is done to save you money and lower your home energy needs throughout the winter season.

The EPA estimates that home energy costs amount to forty percent of the operating budget for a single family home.  Finding ways to lower your demand for heating and electricity can help reduce demand on power plants that use fossil fuels, making it great for your wallet and for the health of the environment. The nice thing about having a home evaluation is that it will not cost you a lot of money to fix some of the issues in energy efficiencies.  Oftentimes caulk and weather sealing is sufficient in lowering your home energy bills.  Changing out all of your light bulbs to compact fluorescents also saves you money; plus, compact fluorescents last longer and have lesser to no mercury, which can end up in our landfills.  Changing the air filters in your HVAC during the upcoming months is also crucial in keeping your systems running efficiently.

Fall clean up around the house can bring about an improved mental outlook; getting rid of junk in and around the house opens up more space and brings with it a better perspective.  It can also make you reconsider the consumer consumption mentality that affects us all.  Uncluttering your home unclutters your life. Trust me on this one—I just did it and feel a whole lot better.

## Green Jobs Equals Green Growth

Green clean tech jobs have become one of the bright spots of our economic environment. A report issued by the research firm, Clean Edge, had the green technology sector showing the biggest gains in employment. The report broke down the top five clean green jobs and the top fifteen markets for those jobs. The top five job areas showing the highest growth were in solar, biofuel, conservation, smart grids and efficiency, and advanced vehicles. The geographic regions showing the greatest gains in job creation were found in the San Francisco-Oakland-San Jose area; Los Angeles metro market, including Orange County; New York metro, including Northern New Jersey and Long Island; Boston metro; and Washington D.C-Baltimore-Northern Virginia. The Chicago metro area came in ninth out of the total top fifteen regional markets. Clean Edge's research report based its findings on the number of job postings and placements, and public and private investments in different economic sectors. One of the interesting highlights was their comparison to the past high tech growth phenomena in job creation; the report showed that previous high tech growth was fairly isolated within the US, hitting only a few of the metro markets.

Green clean tech growth is almost universal in its job growth and investment placements; jobs in this sector are being created in a much broader environment than what high tech offered. Given the current economic environment, it makes sense for government and private equity groups to invest heavily in green technology sectors. Green jobs will provide the greatest opportunities for the unemployed and for future generations of college graduates. The best part of this green job growth is that it can be applied to every community in America. The faster we invest in green jobs, the faster we can solve our dependencies on foreign oil and start the process of ending our past contributions to global warming[117].

**Light Pollution**

Light pollution is that awful orange glow you often see from cities at night. Astronomers both professional and amateur would love to see all of the lights turned off completely. I have to agree with them on that one. If you live in an urban area, you rarely see stars out at night, except perhaps the North Star and a few constellations; on the other hand, if you live in areas far from light sources, you get to see a night filled with stars.

Light pollution is a wasteful excess of modern day society. Do all large corporate retailers need to have lights on throughout the night? Does every street in America need to be lit up? Are we that afraid of the dark? This excessiveness extends to energy and dollars, too. The amount of oil and fossil fuels expended to light up our cities is wasteful and costly. If we learned to turn off lights, we could make a serious dent in global warming and $CO_2$ production. It would also bring down the cost of our fuel bills; some estimates suggest that the amount of money saved would be as high as $10 billion if we just turned off a few more lights[118].

There do happen to be new light fixture solutions available to help lower that late-night city glow. The light fixtures are IDA certified and significantly reduce the affect of light pollution at night. Motion sensors coupled with IDA certified fixtures are a great place to start. Imagine lights going on as you drive or walk by on the street. We need new ordinances that have businesses and governments turning off their lights after 10 pm. Please no more corporate lights and trashy neon signs—I want to see the stars again at night without having to travel to Monument Valley every time I want to see them.

**Natural Products Expo Booms**

Twice a year companies within the natural health food industry gather for the Natural Products Expo. The springtime Expo is held in Anaheim, California, and seems to be a harbinger for seasonal change. Having just finished attending the show, I found that it marked a significant departure from previous shows I've attended. Over the last several years, the state of economic well-being slowly eroded the natural health industry, which was once considered the media darling for explosive growth and profitability. Still, during the 2010 Expo, the halls were crowded in a manner I had not seen in years.

If you ever get the chance to visit such a show I highly recommend going. The Natural Products Expo exhibitors provide countless free samples of food items, personal care products, supplements, and household products. If you are interested in natural products, the exhibitors in attendance will remind you of shopping in a Whole Foods store. Every label you support is most likely at the show.

The best news that came out of the 2010 Expo was that first quarter sales for everyone there seemed to have picked up over the previous year, which is great news for a future economic recovery in the green industry. It's also great news for the rest of us.

# Section Three: For the Future

"Let every individual and institution now think and act as a responsible trustee of the Earth, seeking choices in ecology, economics and ethics that will provide a sustainable future, eliminate pollution, poverty and violence, awaken the wonder of life and foster peaceful progress in the human adventure."

- John McConnell, *found of International Earth Day*

-

## Chapter Ten: Energy

"Ultimately I believe—because energy is so central to our lives—that a common global project to rewire the world with clean energy could be the first step on a path to global peace and global democracy—even in today's deeply troubled world."

- Ross Gelbspan, *journalist, author and environmentalist*

**Light Tubes**

Light tubes were first used in ancient Egypt by the Pharos, who used them to bring light into the pyramids. Light tubes are used today to transport and distribute natural and artificial light and are one of the most energy efficient means of providing sunlight to homes and buildings.

The efficiency of a light tube far exceeds that of a skylight. Skylights in homes and buildings impact energy usage. To heat and cool a building with multiple skylights requires more energy than a building using light tubes. The amount of sunlight that enters skylights causes buildings to warm up during the summer, forcing air conditioning systems to work harder to cool the buildings down; conversely, in the winter, heat rises through skylights, forcing heating systems to work harder. Light tubes instead reflect and intensify sunlight into buildings, providing energy-free light sources that don't alter heating or cooling.

Light tubes have built-in reflectors that magnify incoming sunlight through the tube, similar to a periscope, and provide enough light to replace a fluorescence-based alternative. They can be straight or angled, but a shorter and straighter tube provides the maximum amount of light to pass through. A diffuser at the end of the light tube then disperses the light into a room.

Light tubes are currently being used in homes, factories, and subterranean train stations. Buildings that utilize skylights should be provided an incentive to replace them with light tubes. This is something that should be mandated on future building constructions around the world.

**Kinetic Energy**

Kinetic energy is what a body possesses by virtue of being in motion: the energy is captured and stored until enough of it has been gathered. This energy is now finding its place in an unusual environment.

In Hillside, NJ, at the local Burger King, they are installing kinetic energy devices that will rotate every time a car rolls through the drive-thru lane. The Hillside restaurant manager states that 150,000 cars pass through their drive-thru lane each year. Installation of the energy device is a test concept that the Burger King Company is assessing for potential use in all of their drive-thru restaurants.

The kinetic energy idea is a good one; it's novel and innovative. At least, it's a start in the right direction[119].

**Wave Power**

One of the most promising, yet least mentioned, renewable energy sources is wave power. As you may or may not know, tidal waves have the ability to provide us with additional pollution-free sustainable energy.

There are currently three methods of producing energy from waves that have not been commercially exploited. One of the possible methods is the use of wind stations. Wind stations allow water to enter a chamber and push air up within that chamber, which then powers a turbine. This solution is placed on land near a coast. Another possible method is in the use of a device called a Pelamis. The Pelamis is a land-based device, about the length of five railroad cars, that extends into the ocean and bobs up and down on the waves. The motion drives hydraulic fluid, which then drives a turbine. The last device, CETO, is placed on the bottom of the ocean floor. As the water moves over it, the device sends power to a land-based generator.

Right now, we need to focus on every possible sustainable energy platform that is available to us. Wave power is not the final solution; however, based on how much of the Earth is covered by water, it should be studied and implemented more extensively.

## Geothermal Heating

Since the Paleolithic era, man has been using geothermal heat to keep warm. Hot springs have provided warmth and hot water during the winter months in all parts of the world. Swedish spas, for instance, have been powered through geothermal heat. Even Iceland uses geothermal piping under its streets to help melt the snow during the winter.

Right now, less than one percent of the world's heating comes from geothermal sources; yet, it is a growing clean energy industry within the US. Since the economic stimulus package passed, an additional government tax incentive of up to thirty percent was added to offset the cost of installation. Now is the time to look into these renewable energy sources.

Heat sources in the ground are derived from the natural radioactive decay of minerals and from solar energy absorbed by the ground. Heat can also be derived from geothermal wells close to the tectonic plate border in the Earth. From that well, geothermal pumps can extract the heat in order to provide heat and hot water in the winter months. In the summer, they pump hot air out of one's home back into the ground, keeping the home cooler.

The cost of a geothermal heating system runs around $30,000[120] for an average home. It is a much higher cost than a normal HVAC system, but with government incentives and no heating bill to contend with, the payoff is closer than you think.

**Earth Hour**

Recently, my daughter went around our home turning off lights and announcing it was time for Earth Hour. We lit a few candles and enjoyed the next hour in candlelight and artful conversation— a welcomed departure from video games, TV, and channel surfing. I checked my neighbors to see if they were doing the same, but alas, it seemed that they were not aware of the event.

Earth Hour is a great concept and one that should be extended not just to one hour a year, but to everyday. Last year, our local electric company stated that the Earth Hour event in the Chicagoland area reduced electricity demand by five percent and saved over 840,000 pounds of carbon dioxide emissions. Why not just turn off the power globally? Then we would get to see what the world was like before the electric lightbulb. Astronomers would go nuts with access to a light-free sky. Imagine what this would do to reduce our global emissions. We may even slow down a bit and get back to having salon-style conversations again; then we might invite the neighbors over.[121]

## Green Stimulus

I've been on the road a lot lately, traveling to cities like Tulsa, Bentonville, and Palm Springs. Along the way, I've seen quite a few interesting things. Most recently, I've been in Palm Springs and have gotten a good look at a wind farm that is located to the west of the city. It is a very large wind farm that extends from one side of a valley to the next. The valley seems to act as a natural wind tunnel and a perfect location to set up a wind farm. From a purely aesthetic point of view, it looks quite magnificent. It gets you thinking about why we're not setting up more wind farms in other parts of the country.

The Obama team is talking about creating more green jobs. Well, the obvious was staring me in the face. Why doesn't our economic stimulus bill contain provisions for building massive wind farms around the country? Instead of spending loads of money on roads and bridges, why not invest in ways to reduce our dependency on foreign oil and energy?

We need a green stimulus that spends money on building wind farms throughout the US; and we need to provide massive investments in building cheap solar panels, as well as financial incentives to convert people's homes to solar. Think what that would do to the scales of economy in manufacturing solar panels and wind turbines, not to mention the hiring that would ensue in building and installing clean, self-reliant energy sources.

Along with wind and solar, let's also add geothermal energy. More and more people are drilling geothermal wells to tap into the earth's internal heat to heat their homes. Some friends of mine in New Jersey are doing this on an island five miles from the mainland. Their plan is to have their home heated from geothermal heat and covered with solar panels to offset their electrical needs. They're getting off the grid. To me, that's forward thinking. If we're going to have real change in this country, we need that green stimulus bill now.

## Peer Pressure Combats High Electric Bills

Remember peer pressure in high school? There is a new company that recently discovered that peer pressure still works on pretty much everyone—at least when it comes to electrical bills.

The startup company, OPower, learned that by sending consumers a personalized report on their energy consumption in comparison to their neighbors, it could persuade them to save energy. OPower sends utility consumers envelopes with smiley faces if they have reduced their energy consumption. For consumers who increased their consumption, the company sends them a report comparing their behaviors to those of their less energy-consuming counterparts.

OPower discovered that the smiley-faced consumers liked the added attention of being singled out as energy misers. The company also noted that the other consumers reduced their energy use from 2.8 percent to 1.2 percent upon receiving their report. It might not sound like a lot, but that dip is getting the attention of utility companies across the country.

Utility companies are being provided incentives from state governments to lower their customers' energy consumption; getting customers to lower their consumption actually helps the companies. Not surprisingly, utility companies are now getting on the OPower bandwagon. So far, twenty-four companies from around the US have signed up for the service. By combining their data analysis with smart grid technology, OPower predicts that utility companies will be able to help people conserve energy and save money by actions such as replacing inefficient refrigerators, unplugging big-screen TVs, and turning down air conditioners at night.

OPowers efforts are pragmatic in their approach. They provide the least capital-intensive investment in lowering energy consumption using nothing more than peer pressure. Who would have thought that we'd still be playing by high school rules?[122]

## America's Next Energy Revolution

It only takes one idea, one product to change the world: the light bulb, the telephone, the car, the plane, the computer—ok, you get the idea. Recently, *60 Minutes* ran a segment about a concept that has the potential to become that next product. The concept has taken shape in what is being called the Bloom Box: the first fuel cell device that provides grid-free electricity to your home. The Bloom Box itself is cube-shaped and roughly the size of a 6" by 6" by 6" box. According to K.R. Sridhar, the CEO of Bloom Energy and the rocket scientist behind the concept, one Bloom Box can power four to six homes in Asia and one home in Europe (you would need two boxes to power one typical American home.) Sridhar indicated that the box transmits electricity wirelessly to the home and can be powered from sustainable sources such as wind, solar, biomass, and fossil fuels like natural gas. The Bloom Box is cleaner than the current transmission of electricity and could eventually replace the transmission lines found all across the US. The planned cost for a home unit could be as low as $3,000.

K.R. Sridhar's initial idea was to build a box for NASA that could create oxygen on Mars. NASA never tested the box, though, so Sridhar decided to reverse-engineer the box to create a new fuel cell device. Now a commercial product, large Bloom Boxes have been installed and tested at Google, eBay, FedEx, Staples, and Wal-Mart facilities. EBay even went on record to say that the box has provided cost savings on their electric bills and is more efficient in providing electricity than the five acres of solar panels they have on their campus' roofs.

Bloom has a long way to go before they can make the Bloom Box into a viable consumer product, but they are making progress. I wish them all success on making the Bloom Box a viable alternative for the world.[123]

## Navajo Nation Moving Away From Coal

The largest Native American tribe in the US is the Navajo. The Navajo Nation has over 300,000 members and the biggest reservation in the country. One of their economic lifelines has been coal power and mining. Decades of coal have sullied their waters and air; but today there is a movement to move away from coal and embrace solar and wind energies.

Coal has been declining in the Navajo Nation and changes in clean air regulations are forcing the Navajo to consider changes to coal power plants. The EPA and the State of Arizona are both seeking to add more expensive pollution controls to older coal power plants based on Navajo lands. On top of that, states like California, which depends on coal power plants, are increasingly imposing greenhouse gas emission standards; they're also requiring renewable energy purchases and banning or restricting the use of coal for electricity. In response to these changes, the Navajo are now investing in alternative energy solutions. In 2010, the Nation approved a wind farm to be built outside of Flagstaff, AZ; it is expected to power up to 20,000 homes. The Nation is also bringing solar and wind power to 18,000 homes on the reservation without electricity.

Unemployment runs up to sixty percent on the reservation; coal-fired plants and mining operations were key employers for members of the Nation. With coal's decline in the region, the Nation is hard pressed to create new energy jobs. Their move to solar and wind energies is the right step to take.[124]

## Star Power Here on Earth

Imagine if we could harness the power of a star on planet earth. Could it be the ultimate clean energy scenario for our future? Does it sound like science fiction? Not if you're spending three billion dollars on building a star prototype here on earth. This concept is being explored at the National Ignition Facility, a large-scale experiment in laser fusion based at the Lawrence Livermore National Laboratory. Scientists at NIF are looking at the facility as a potential key to producing large amounts of carbon-free power. Just how much power, you ask? They are trying to produce four million joules, which is enough to produce four million watts of power in a single second. NIF plans on taking a single laser beam, no wider than a human hair, and turning it into 192 beams that would individually be eighteen inches wide. Together, the beams would then produce the four million joules. The NIF will employ a series of mirrors and amplifiers to route the laser beam into multiple beams. All of these laser beams will create some serious heat—as high as a hundred million degrees Celsius. The heat created will provide the carbon-free energy scientists are hoping to obtain from the system. Right now, the scientists can only fire the laser system once every two to three hours. The hope is that eventually they will be able to fire the laser several times per second.

The NIF facility was completed in 2009 and comprises two ten-story buildings. Scientists hope that within fifteen to twenty years they can make the facility into a viable power plant system. Imagine then, we could have a small star housed in Northern California providing clean, sustainable carbon-free energy.[125]

**Osmotic Energy**

The country of Norway plans on using osmotic energy to harness the power of sea salt. Statkraft, a Norwegian-owned renewable energy company, will operate a prototype power plant on the banks of the Oslo fjord. Their goal is to create energy through the contact of freshwater and saltwater through a membrane. Osmotic energy—think Osmosis—is based on the principle that nature is constantly seeking balance and plays on the different concentration levels of liquids. When freshwater and seawater meet on either side of a membrane, a thin layer that retains salt but lets water pass, freshwater is drawn towards the seawater side. The resulting flow puts pressure on that side. That pressure can then be used to drive a turbine, producing electricity. Statkraft was drawn to the possibility of creating energy through this process because of its predictability. As a renewable energy source, it is always constant unlike wind and solar, which can be affected by external weather factors.

Statkraft's prototype power plant will soon be operational. If it is successful, the company plans on building a commercial-grade plant. The prototype plant will produce two to four kilowatts of energy, while the commercial-size plant on the drawing board will generate twenty-five megawatts, which is enough power for 10,000 homes. If the prototype plant is successful, the commercial plant will be operational by 2015. The company estimates that global potential for osmotic energy production could equal 1,700 terawatts—enough power to light up all of Europe. The key to success is the membrane that will be used to have saltwater and freshwater flow through. Current technology allows for three watts of power for every meter of membrane. To make this a profitable venture, efficiencies need to be raised to five watts per meter. If osmotic energy does play out, it will be one more contribution to making renewable energy on this planet.[126]

## Small Wind Turbines Gain

For those who may not know, a small wind turbine is defined as one that can generate 1200 kilowatt-hours a year. The average wind speed for a small turbine is 11.2 miles per hour; the turbine starts generating at seven miles per hour. Small wind turbine sales have surged in the last several years, with 10,000 new units installed throughout the country. The small turbines—defined as 100-kilowatt capacity or less—grew fifteen percent in 2009, representing $82.4 million in sales and almost 10,000 new units, according to the American Wind Energy Association (AWEA). Sales were aided by a thirty percent federal tax credit for renewable energy investments and state incentives.

Maximizing the efficiency of a wind turbine through proper placement optimizes the amount of electricity that is generated. One of the obstacles in installing small wind turbines is determining where to place them on your property. To aid in that process, the state of Massachusetts has devised an online software program to help homeowners and businesses determine the best location for their turbines. Massachusetts offers a rebate based on ongoing performance of a wind turbine, proper placement can increase one's rebate from that state.

In the past, installers and homeowners relied on satellite data to get an idea of available wind resources. New tools in assessing wind flow have helped improve wind turbine deployment and have helped increase the sales of small wind turbines. There are several noteworthy small wind turbine manufacturers to choose from including Swift Turbine, Helix, Windtronics, and Southwest Windpower. Helping homeowners and businesses improve their site selection only strengthens alternative energy's contribution to dumping fossil fuels.[127]

## Worldwide Gains in Sustainable Energy

According to the UN's International Energy Agency, in 2009 more than half of all new electricity added in the United States and Europe came from renewable power sources such as wind and solar. Also, 2009 was a record year for the amount of new green power added to the grid, partly a result of shifting deployment of manufacturing away from flagging developed countries to emerging economies, including Brazil, India, and China.

The IEA issued their annual report, the Renewable Energy Policy Network for the 21$^{st}$ Century (REN21). Of the extra eighty gigawatts of new renewable power capacity added worldwide, China added thirty-seven GW, more than any other country, according to the report. Despite the impact of the financial crisis and lower oil prices, renewable capacity grew at rates close to those in previous years, including solar photovoltaic power at fifty-three percent and wind power at thirty-two percent. Based on an article written for CNET News, grid-connected solar photovoltaic power had grown by an average of sixty percent per year for the past decade, increasing a hundred fold since 2000. The boom has been largely on the back of support in European countries, where a pullback following the recession raised investor jitters; but the wind and solar sectors are still poised for a record year in 2010, operators and investors predict. While China is making great strides in renewable energy deployment, its carbon emissions also accelerated in 2009, placing it even further ahead as the world's top emitter of the main greenhouse gas blamed for climate change.

Wind and solar power together make up less than three percent of the total US power generation, though both are growing rapidly amid a range of state and federal incentives. The fact that fifty percent of all new power plants are coming online both nationally and internationally illustrates the strides we are making in building sustainable energy sources. As each year passes, the hope is that we can eventually raise that number to a hundred percent.[128]

## Chapter Eleven: Regulations

"Sometimes it's not enough that we do our best; sometimes we have to do what's required."

- Sir Winston Churchill, *Former Prime Minister of the United Kingdom*

**Natural Capital**

We all know what financial capital is in economic terms, since it's been one of the main conversations we've had around the company water cooler; but, a type of capital that needs our undivided attention is something called Natural Capital. Already, society has been discussing natural capital without really calling it that.

Natural Capital encompasses intact forests, healthy air, clean water, plant and animal biodiversity, and a stable climate. I would venture to say that we have not placed an economic and social cost on our natural capital. Unlike governments that exercise poor financial planning and then compensate by printing new money, natural capital has no resources to draw upon for replenishment; particularly when man is involved. We can't "print" fresh air and clean water; we have to work to fix what we have because at the end of the day, it is finite.

In order to protect our economic and social welfare, we have to protect our natural capital; it needs to be our primary focus. Spending economic stimulus for the creation and protection of natural capital should have been our first investment. The sooner we wake up to that fact, the sooner we can get our natural capital financial house in order.

## Design for the Environment

Design for the environment has nothing to do with walking down the catwalk in Milan. Instead, it's an initiative by the US Environmental Protection Agency for industry and manufacturers. The agency's initiative is to steer different industries in the direction of creating products that are less harmful to people and that lessen the impact of pollution on the environment. The downside of the initiative is that it requires companies to spend a lot of money to prove that the products they sell meet EPA guidelines. It may be important for companies that don't actually make their own products, but for guys like us who make everything ourselves and base our business on being green, we end up getting taken to the cleaners.

Unfortunately, there has been a lot of "greenwashing" in the natural product category, so outside scrutiny is not a bad thing. The one downside to the EPA's initiative, though, is that it isn't pushing industry as hard as it could; what they're seeking is a slow evolutionary process over time to get companies into compliance. The compliance right now is voluntary, but it needs to be mandatory.

The initial product testing for the EPA's initiative takes months and the usual government bureaucracy takes just as long. The EPA is moving too slowly on targeting household cleaners that contain formaldehyde, 1,4-dioxane, chlorine, phosphates, and ammonia; these are ingredients that we as consumers need to be more concerned about. For more information on this program, visit *www.epa.gov/def/*.

## Phosphates are still an Issue

In 1974, when the US EPA was created, one of their first efforts at regulating pollution was a phosphate ban on laundry detergents. Phosphates found in laundry detergents had a devastating effect on aquatic life, while wastewater treatment plants had a difficult time eradicating phosphates from their systems. Within a short time, all laundry detergents had removed phosphates from their formulas.

The problem with phosphates is that they are still used in automatic dishwashing detergents and other household cleaners. The City of Chicago banned phosphates in auto dish detergents, but has never enforced the ban. Their claim has been that no detergents exist without phosphates; but that's not actually true. Just look to our Wave Auto Dishwashing detergent, which is produced here in Illinois. I even testified at the state capitol in Illinois that our company did indeed have a phosphate-free formula and that a total ban on phosphates would not impact consumers. Four counties in Washington State have already banned the use of phosphates in all products; Minnesota, Illinois, New York, and California have similar legislation, which will soon go into effect.

It's amazing that after thirty-five years our government cannot enforce a total phosphate ban in household products. Maybe this is one more change that can occur in Washington.

## 1,4 – Dioxane – What's it all About?

I'm not sure if you've heard or are even aware of the controversy surrounding 1,4 Dioxane in personal care and household products. The subject of 1,4 Dioxane was raised by the Organic Consumers Association at one of the trade shows we annually attend. The Association had tested a variety of personal and household cleaning products produced by several leading manufacturers. The testing revealed that 1,4-Dioxane was in quite a few products from a variety of brands found in natural health food stores—our Dishmate Dishwashing Liquid was one of them.

For those of you who are unfamiliar with 1,4 Dioxane, it is a byproduct of the way certain cleaning agents in household products are produced. These cleaning agents are called surfactants. As it turned out, one of the surfactants we were using had trace elements of 1,4 Dioxane. We were unaware of this in our Dishmate and had been under the impression from our supplier that it was not present. When OCA brought this to our attention, we were mortified; we quickly moved to eliminate it from the product. Still, the whole issue got us checking everything we produced from top to bottom all over again.

Now the question I haven't answered (for those of you unsure what I'm talking about), is why 1,4 Dioxane is a big deal. Well, studies have found that products containing high levels of 1,4 Dioxane can cause cancer in laboratory mice. The State of California has enacted legislation that bans levels of 1,4 Dioxane above a certain threshold and the US EPA is studying whether to do the same on a national level. The moral of this story is to make sure that the personal care and household cleaning items you buy do not contain this ingredient.

The OCA has a website, *www.organicconsumers.org*, that shows their test results; but if you really want to dive in deeper, the Internet has quite a few articles on this issue.

## New CAFE Standards in Effect

The New Corporate Average Fuel Economy standards from the Federal Government recently went into effect, usurping all state requirements in setting fuel efficiencies. The standards require automakers to raise the average fuel efficiency of their vehicles from 27.5 mpg to 34.1 mpg by the 2016 model year. The new standards are expected to reduce $CO_2$ emissions by roughly thirty percent between 2012 and 2016, and save the country $240 billion in fuel savings. They are also expected to reduce pollution and possibly imports.

So, what can we as consumers expect from new fuel economy standards? Like with all new things, we can expect a higher price tag on vehicles. Based on the automakers' expected cost of $52 billion to implement the new standards, Consumer Reports estimates that it will cost consumers an extra $1,100 per vehicle. On the bright side, they expect fuel savings to increase significantly: up to $3,000 over the life of the car.

New cars are expected to be smaller and lighter, and automakers are expected to herald in more hybrids and all-electrics. We may even see new dual-clutch and seven or eight speed automatic transmissions that can increase fuel efficiencies; even more diesel vehicles could become the norm to help boost fuel economy.

The new CAFE standards are focused primarily on improving gasoline engines to help increase fuel efficiency. Hopefully, the new innovations to boost miles per gallon will result in some exotic new looks to the cars we drive. Perhaps a revolutionary design to reduce airflow could provide us with that stealth car we all dream of driving one day.[129]

## Obama Stalls Animal Antibiotics Act

New York Congresswoman Louise Slaughter has a bill pending in the House Rules Committee that would restrict antibiotic use in food animals. Rep. Slaughter speaks from experience on this pending legislation—she is the only microbiologist in Congress. Scientists and doctors who fear that the overuse of these drugs makes them less effective in fighting bacteria join her. The Obama Administration is currently stalling the bill—the same administration that had originally condemned the practice of feeding antibiotics to cattle, hogs, and poultry.[130]

The Union of Concerned Scientists estimates that as much as seventy percent of the antibiotics used in the United States are given to healthy animals. Conventional farmers and ranchers routinely feed antibiotics to their herd to help the animals use their food more efficiently and bulk up faster. They say the medication also helps ward off pathogens that could sicken or kill their livestock; but even the US Food and Drug Administration is concerned about giving antibiotics to food animals. "We're looking at ways to phase out the use of antibiotics for growth promotion and food efficiency in livestock," said Dr. Joshua Sharfstein, principal deputy commissioner of the US Food and Drug Administration. Dr. Sharfstein is concerned that giving anti-microbials to healthy animals could contribute to more drug-resistant infections in people. Sharfstein is also pushing for veterinarians to oversee antibiotic use on animals.

Currently, livestock feeds mixed with antibiotics such as Tylosin, a macrolide, and Chlortetracycline are sold over the counter. The deputy commissioner recently submitted his written position to the House Rules Committee on the Preservation of Antibiotics for Medical Treatment Act. According to Rep. Slaughter, one of the reasons meat and poultry producers use antibiotics is because they house their animals in poor living conditions. For animals living in close proximity to one another, the use of antibiotics is essential to keep them healthy.

This legislation could drastically change farm practices in this country. Many consumers have already turned to antibiotic-free meat and poultry because they want products that have been raised naturally and from industrial farm settings. We need President Obama to be that catalyst for change he had promised to be.[131]

**Twenty-Four Indicators of Global Warming**

Hold on to your hats everyone. The EPA, after a decade of inactivity and denial, finally released a report acknowledging that the climate is indeed changing. The EPA under the current administration released a report titled "Climate Change Indicators in the United States," which measured twenty-four separate indicators that climate change is affecting the health and welfare of the country. I've included several of the key climate indicators below:

- Greenhouse gas emissions from human sources are increasing; emissions in the US have grown by fourteen percent from 1990 to 2008.

- Average temperatures are rising; seven of the ten warmest years on record have occurred since 1990.

- The intensity of tropical cyclones has increased in recent decades. Of the ten most active hurricane seasons, six have happened since the mid-1990's.

- Between 1993 and 2008, sea levels have risen at twice the rate of the long-term trend.

- Glaciers are melting and their loss of volume has accelerated over the last decade.

- The frequency of heat waves has steadily risen since the 1960's; the percentage of the US population experiencing heat waves has also increased.

Collecting and analyzing environmental indicators can help us understand the causes of climate change, as well as predict what the future might bring. This information is critical so that we can devise strategies for avoiding the worst effects of climate change and for adapting to a different climate.

The EPA's report primarily describes trends within the United States but also includes global trends to provide a basis for comparison.  The report includes some sobering statistics about how climate change is affecting temperature, precipitation, sea levels, and extreme weather.  The EPA has taken the first step in helping Congress pass the Climate Energy Bill; and it couldn't have come at a more opportune time.[132]

## Phosphate Ban Gaining

In July 2010, sixteen states enacted a total phosphate ban on auto dishwashing detergents. The bill was primarily targeted at dishwashing detergent products, since the last legislation in the 1970's specifically targeted phosphates in laundry detergents. The Soap and Detergent Association, an industry-lobbying group, provided a synopsis of the legislative activity around the ban. The majority of states that participated in banning phosphates passed legislation in 2007 and gave the industry three years in which to find an alternative. The sixteen states that passed the legislation included Illinois, Indiana, Maryland, Massachusetts, Michigan, Montana, New Hampshire, Ohio, Oregon, Pennsylvania, Utah, Vermont, Virginia, Washington, and Wisconsin. The only state that vetoed a total ban on phosphates was California. All states had an effective date of July 1, 2010 for a reduction in the use of phosphorus in household automatic dishwashing detergents to a maximum of 0.5 percent (by weight). Unfortunately, the ban did not extend to commercial dishwashing applications, which meant that every hospital, industrial complex, restaurant, school, and public facility could continue to use detergents with phosphates.

What makes phosphates such a problem is their environmental impact; phosphates can cause excess algae blooms in our waterways, which depletes the oxygen available to aquatic wildlife. There are a few companies right now, though, that are making phosphate-free auto dishwashing detergents, including Earth Friendly Products. Earth Friendly Products' Wave Dishwashing Gel was the first pH neutral, phosphate-free product in the marketplace.

State governments need to push the envelope here and provide their citizens with a total phosphate ban. Leaving out commercial use of phosphates is not the right thing to do. We need a complete ban and we need it now.[133]

**American Power Act Revealed**

Senators John Kerry and Joe Lieberman are sponsoring the American Power Act legislation, an energy bill that they recently released to the public. The legislation aims to cut emissions of carbon dioxide and other heat-trapping greenhouse gases by seventeen percent by 2020 and by more than eighty percent by 2050. For the first time, the legislation would set a price on carbon emissions for large polluters such as coal-fired power plants; depending on market prices, rates initially would range from $12 per ton of carbon emissions to $25 per ton. Restrictions would not take effect for power plants and transportation fuels until 2013 and 2016 for manufacturers. Allowances would be granted to local electrical companies, who would be required to use them to help ratepayers. In addition, a separate consumer relief provision would provide rebates to eligible families.

Senators Kerry and Lieberman said that the bill would exempt farms and most small and medium-sized businesses and instead concentrate efforts on the largest polluters. The bill would also allow coastal states to opt out of drilling, being allowed up to seventy-five miles from their shorelines—a concession to lawmakers concerned about offshore drilling in the wake of the Gulf Coast oil spill. In a break from current policy, states that allow offshore drilling will receive a share of federal revenue, the summary states. That provision is likely to spark debate from interior senators, mostly in the West, who object to revenue sharing for offshore drilling. Kerry and Lieberman have said they will press ahead with the climate bill despite losing the support of their only Republican partner, Sen. Lindsey Graham, R-S.C.[134]

## $8 Billion for Midwest High Speed Rail

President Obama recently earmarked $8 billion for a high-speed Midwest rail service. The funds will be used to link the cities of Chicago, St. Louis, Minneapolis, Detroit, Milwaukee, and Indianapolis. The rail service is expected to hit land speeds of 110 mph to each of those cities from Chicago, providing for a faster travel time than driving. It looks like the romance of high-speed rails may soon be upon us.

Funding for the rail service was being drawn from the Obama Administration's economic stimulus plan, with implementation of the new service expected by 2013. The service links the cities to each other through Chicago. Details of the type of service that will be provided have yet to be determined; but the states and cities involved have been working on such a master plan for some time now.

Most of the engineering requirements for the service have already been completed, including the designation of the physical routes through each state. Now, if we could only aspire to create a national bullet train system that matches the Europeans, the Japanese, and the Chinese.[135]

## It's 1984—Corporate Style

The US Supreme Court ruled that corporate America, unions, and lobbyists are free to spend as much money as they want on political candidates; money is apparently no longer an object during campaigns. Now, any efforts to limit this kind of influence have been crushed by the Supreme Court. It makes you wonder how much money was sent their way in order to rule in this manner.

Congress and the White House have announced that they will try to reformulate new campaign spending limits; but I view that as the fox guarding the chicken coop. We as a nation of uninterested voters seem not to care that we've lost control of our own country. Less than fifty percent of registered voters turn out to vote in elections, except perhaps in the 2008 presidential race (you remember that race; the one that was supposed to change everything). Corporate and Union interests understand this, as do our politicians.

It seems that we've given up on our American experiment. Large vested interests have taken control over governance, similar to the totalitarian story found in the book *1984*. Our government workers, who we pay to look out for our interests, are now firmly in the pockets of corporate and union benefactors. What will it take for a country built on the sweat and blood of our forefathers to wake up and take back their country?

Thomas Jefferson once argued that for a country to stay true to the will of the people, there should be a generational revolution once every twenty years. He felt that the moneyed interests and the established politicians would not be receptive to the interests of the people. He knew what we would face, for every society has experienced the same phenomena. Revolution is not a practical formula for a society to go through every twenty years. The institutions created by our forefathers provide us with the ability to change things. We need to become better, more informed, and more active citizens; we should not fear our government and those who have the means to control it.[136]

**EPA Finally Acting Like the EPA**

The science has finally returned to the US Environmental Protection Agency. The EPA announced that, under the Clean Air Act, they have the authority to regulate greenhouse gases. The gases covered under the Clean Air Act include carbon dioxide, nitrous oxide, methane, perfluorocarbons, hydrofluorocarbons, and sulfur hexafluoride. The EPA found that the projected concentrations of these greenhouse gases threaten the public health and welfare of current and future generations. The combined gases come from mobile sources (all motor vehicles), and stationary sources (power plants and industrial manufacturing).

The impetus behind the EPA's announcement came from two sources. The first was a lawsuit that ended up in the Supreme Court, Massachusetts v. EPA. The court ruled that the EPA has the authority to regulate greenhouse gases under the Clean Air Act. The second source came from the same court, allowing the EPA Administrator to evaluate whether greenhouse gases endanger the public welfare. The EPA then opened a sixty-day public comment period on their findings that greenhouse gases endanger the public welfare; they received over 380,000 public comments. From these comments, which included scientific research, the EPA concluded that greenhouse gases are affecting rising sea levels, primarily on the East Coast and Gulf Coast, and threaten human populations along those coasts. Water levels and supplies are being threatened, impacting snow packs and seasonal water supplies in the western part of the US. Rising temperatures also increase weather variations, which can cause damage to crop yields and forestry; and ecosystems will have to shift their habitats to higher elevations and move northward to offset the higher temperatures.

Climate change no doubt exacerbates human migratory patterns and threatens the country's security. If the EPA can show a strong political resolve in reducing greenhouse gas output in the United States, then we will be making a huge difference for ourselves and for future generations.[137]

## Lost Trees

Bruce Springsteen occasionally sings the refrain, "lost in the swamps of Jersey". Most people not from New Jersey immediately think of the smoke stacks off the turnpike; they often think of Jersey as an ugly state. That's fine by me, because that means less people at the shore in the summer. Unfortunately, that's not actually the case. In reality, a growing influx of people in the area has prompted the New Jersey Department of Transportation to widen the Garden State Parkway.

The Garden State Parkway is one the prettiest roadways in America—and one of the most congested on summer weekends. The parkway runs through one of the nation's best-kept secrets: the Pinelands National Preserve. The Preserve is one of the country's largest pine forests and lies in the little state of New Jersey. What's even more impressive is that the largest water aquifer system in the US lies below it; it also happens to be one of the last remaining untapped aquifers.

The Pinelands National Preserve is one big forest; it is also a forest under attack from the NJ Department of Transportation. The department is clearing a path through the Pinelands to add more road lanes so as not to inconvenience summer visitors. There have been massive tree clearings that look like something you would see in the Amazon; it is a travesty. Adding more asphalt and vehicles spewing $CO_2$ into the air will not help this ecosystem and the people living near it; the loss of trees will also impact the amount of $CO_2$ in the air. For every tree that comes down—and not just in the Pinelands—there should be a law forcing state departments and companies to plant a new tree in its place. Then it might cause them to pause before they plow a new path in the name of tourism.

## When is Enough Enough?

When do taxpayers say enough is enough? When will we see our local, state, and federal governments find the political backbone to solve their financial irresponsibility? This really isn't about the over-taxation that exists in this nation, but about the lack of fiscal solvency; fiscal solvency doesn't exist in the majority of local, state and federal governments. Let's forget about our massive federal debt for one moment. Every local and state government is in trouble; every politician talks about having the will to put more of the burden on taxpayers, but then decides not to do it during an election year. Are we really that stupid? Are we nothing but sheep being led to slaughter?

I find it distressing that politicians talk about raising our income tax level or raising sales tax or placing another tax on services. None of them discuss putting their own houses in order. What would it take? Well, let's start with the basics. You cannot spend money that you don't have. You can borrow short-term to make ends meet, knowing that down the road the economy will pick up and tax revenues will increase; but what are they to do about it now that money coming in is way short of what's going out? What would you do? Simply cut costs to make ends meet? Would that entail public employee layoffs or eliminating pension funding? How about sharing resources with other government agencies to reduce costs and overhead, maybe simplifying government services so they are not confusing to the public.

Some local governments have begun to pool together their resources, primarily in policing and fire protection, which makes sense. How often is the fire department in your town called upon to put out a fire? In my town, all I see are police cars setting up speed traps to produce revenue. The government needs to put its houses in order and make the painful cuts everyone else has had to endure during recent years. If not, we will eventually have had enough of this and do something that would not make our founding fathers proud.

**Democracy in Action**

I recently had the privilege of being called upon for jury duty. I must admit, it was the first time I actually had to show up for it. Jury duty is something I've always tried to avoid; time, boredom, and over-exposure to too many Perry Mason episodes have influenced my behavior towards the system. I've also felt that there is a conflict of interest within our judicial system. Lawyers can become judges and judges can become lawyers. If one couldn't become the other, I think perhaps then I may change my viewpoint; but, both groups play off one another and help perpetuate their existence and compensation. There isn't necessarily anything wrong with that. It's just the way the system operates. So, as I sat in a room with my fellow citizens, all quietly, (or not so quietly), watching TV or reading, I waited to hear if I'd been selected. Ironically, that's all that happened; we waited and then were told we could leave. Though it was a letdown, I was relieved that I wasn't called upon to serve.

Jury duty is probably the most direct democratic participation we have in this country; no other country allows a jury of peers to decide the outcome of a case. Instead, we as citizens decide the outcome—no government agency or powerful corporation; it's just us. In my view, it's one of the best things we have besides voting. Though jury duty can be a pain in the butt, it's a part of our democracy that I wouldn't give up.

## Obama Ushers in the National Ocean Council

President Obama recently set a new policy to improve coordination and communication between agencies that administer the use of US coastal waters, including the Great Lakes. The newly created National Ocean Council will try to make sense of the various rules regarding the use of US coastal waters and the Great Lakes. While the council won't change any existing laws or regulations, its main focus will be on how best to manage the competing uses of our oceans and Great Lakes. The council will include secretaries of all cabinet-level federal agencies and representatives of other federal environmental and economic agencies. State and tribal authorities will be allowed to comment on the council's future recommendations. The issues addressed will include the ability of ocean and coastal ecosystems to remain resilient and adapt to the effects of climate change, as well as the expected acidification of oceans; promoting and implementing sustainable practices on land, and concerns about water quality will also be addressed.

The policies call for special attention to the Arctic Ocean and adjacent areas expected to face dramatic changes from global warming, including rising sea levels. A key facet of the new national policy will be a reliance on scientific knowledge, which requires greater financial support for scientific research and for the tools needed to conduct that research.

It's unfortunate that it took a devastating oil spill to galvanize the leadership in Washington to wake up and take notice of what's going on in our oceans; but the fact that they have is welcomed. Hopefully the council can develop a compressive response to maintaining this vital resource.[138]

## Biofuels get $600 Million from Exxon Mobil

From the corporation that previously dismissed alternative biofuel ethanol as "moonshine", comes a $600 million joint investment into turning pond algae into biofuels. The joint venture, financed by Exxon Mobil, is with Craig J. Venter's firm, Synthetic Genomics, a biotechnology company. The $600 million project seeks to produce liquid fuels for transportation. Under the partnership, Synthetic Genomics will research and develop systems to grow large amounts of algae and convert it into biofuels. Exxon Mobil is providing engineering and scientific support to increase the levels of algae production in manufacturing the final product. Algae biofuels are already in existence and are used successfully as cooking oil in Nicaragua. Though the concept isn't new, having a company like Exxon Mobil behind it could bolster high yields of biofuels for mass use. Exxon Mobil is the world's largest oil company and sits on the top spots of the Fortune 500. This kind of commitment equaled roughly five percent of their total sales volume for 2008. Hopefully the company will push to get this product to market. It's been done before and we know that it works.[139]

## Chapter Twelve: Transportation

"There's so much pollution in the air now that if it weren't for our lungs there'd be no place to put it all."

-Robert Orben, *American magician and comedy writer*

## Electric Cars Gaining Consumer Interest

A national survey conducted by Consumer Reports asked 1,752 adults about their views towards electric cars. The survey concluded that as many as twenty-five percent of US consumers would consider purchasing an electric car as their next vehicle. Although that twenty-five percent would consider buying an electric car, the survey indicated that none of the consumers were keen on the idea of giving up performance or convenience.

Consumer Reports found that the biggest obstacle to electric vehicle acceptance was the vehicle's driving range. Based on the survey's findings, the median driving range consumers would accept for an all-electric car was eight-nine miles; some consumers felt that a seventy-five mile or even a forty-nine mile range would be sufficient for their daily driving needs. Another major factor that played into buying an all-electric car was price. The median added cost consumers said they would pay for an electric was $2,068. Twenty percent said they wouldn't pay anything more, while another twenty percent said they were willing to pay as much as $5,000 extra. One could interpret the top-line result in a different way: a majority of respondents—nearly three quarters—said they didn't expect to consider a plug-in electric car as their next purchase. Other studies, such as one conducted by Ernest & Young, indicated that consumers are indeed interested in electric cars because of their potential to save money on fuel and because they cause less pollution than gas-only vehicles.

Consumers' need for longer driving ranges and availability of charging stations will have an impact on the future success of electric cars. People want choice and convenience; and right now both of those attributes are non-existent.[140]

## Road America

Did you know that the US has more paved roads than any other country in the world? The current tally is up to 2,425,000 miles of paved roads. As of now, choices in road construction consist blacktop and concrete, though the majority of roads in the US are blacktop. Both types of pavement have their pros and cons; and both types have an impact on global warming.

Concrete costs more to install than blacktop, but if it's installed correctly, it can last for decades; most major inner city highways are built with concrete for that reason alone. The longevity of concrete prevents road crews from needing to make frequent repairs, and saves society money and time. Instead of acting as a heat sink, concrete reflects sunlight from heating up the ground; asphalt, on the other hand, absorbs and creates heat. Asphalt comes from petroleum, a finite resource, and when produced, adds to the CO2 levels in the atmosphere. The heat it absorbs contributes to global warming.[141]

Scientists at the University of California estimate that if all road surfaces reflected sunlight, like with concrete, we could offset forty-four metric gigatons of greenhouse gases, which is more than what the entire human population produces annually (the university's estimate also included altering every rooftop to a white one.) That estimate, though, would require global participation in moving from asphalt to concrete; but implementing both changes on a global scale, we could make an easy adjustment that would have an immediate impact on global warming.

## Green Aviation

Continental Airlines, Virgin Airlines, and Japan Airlines recently conducted test flights using biofuels. Each airline ran one engine of a two-engine aircraft on biofuels from different derivatives. Continental Airlines ran on an algae mixture, while Virgin used a hybrid conventional jet fuel made from coconut and babassu nuts; Japan airlines ran its engine on Camelina, a weedy flower from Europe. The flights lasted forty to ninety minutes, possibly proving the viability of biofuels in aviation.[142]

The International Air Transport Association mandated that by 2020 their members would achieve neutral carbon growth (meaning that they would not exceed current CO2 levels). The mandate requires a 1.5 percent annual reduction in CO2 output—even as air travel is expected to increase once the global recession ends. To achieve this number, their airlines will need to move to more fuel-efficient aircrafts and use a higher percentage of biofuels. The aviation industry currently uses over sixty billion gallons of jet fuel on an annual basis. The greatest challenge for the industry, and for the auto industry as well, will be to meet these fuel demands using alternative fuel sources.

Just how practical is the use of biofuels? Unless we can find ways to increase the yields from algae and provide them at a competitive cost, biofuels won't be of much help. We also have to consider the cost of converting forests to agricultural lands in order to increase the yield demands for biofuels. If biofuels are to meet consumer demands, we have to re-examine where we'll grow them and just how much tillable land we're willing to hand over for these future fuel sources.[143]

**Green Deliveries**

Besides the media's focus on green cars and green buses, an often-overlooked development has been the greening of commercial trucks. I'm referring not so much to large highway rigs as to local delivery trucks used by companies like FedEx and UPS. Hybrid technologies don't help out the mileage factor for long haul trucks due to their lack of start and stop driving, which is what boosts hybrid performance; but an initial slow acceptance of using hybrid technologies in delivery trucks is rapidly growing.

Azure Dynamics, a company that moved from Canada to Detroit, is taking the Ford truck chassis and converting them to hybrids using their proprietary hybrid technology. Both FedEx and UPS have gone from prototype testing to integrating these vehicles into their fleets throughout the country. The gains for both companies are significant: hybrid trucks improve fuel economy by forty-two percent, reduce greenhouse gas emissions by approximately thirty percent and cut particulate pollution by ninety-six percent. Not only have these companies gained on their bottom line, they have made a strong contribution to the environment. Another reason to say thanks to your local delivery guy![144]

**Congestion and CO2**

I'm not sure if you've noticed, but recent government reports have indicated that traffic congestion and CO2 levels have gone down. A governmental report on air quality recently stated that CO2 levels are down four percent over 2009 levels. I'm not sure if I'm breathing cleaner air than before, so I'll just take their word on it.

Local governments are reporting an easing of traffic congestion in major cities. Here in Chicago, a new report was issued stating that traffic congestion has also been lowered by as much as four percent; that I've noticed, but not by much. All of the reports cite the recession as a main reason for lower numbers and expect the numbers to climb again once the economy is back on its feet. Even though congestion and pollution are down according to the government, it would be nice to see new reports on how they could work to keep it that way.

## Hybrid Buses

We've all been stuck driving behind a slow-moving bus at some point in our lives; I'm not sure which is worse, being stuck behind a school bus or a public transit one.  Either way, as soon as they stop and start up again the belching diesel exhaust has you gasping for air and scrambling to shut your car windows.  Buses are probably the vehicles I have the least love for; but they are an important part of public transit systems throughout the world.  I've seen prototype buses that run on natural gas, though they can be just as foul smelling as the buses currently on the roads.  An all-electric bus seems out of the question for now, but you never know.

Now there are hybrid buses.  These new buses generate electric power every time they use the break pedal and are able to reduce $CO_2$ emissions by thirty to forty percent over current buses.  They still run on diesel, but their new electric systems aid in acceleration and cruising similar to that of a Toyota Prius or Ford Escape Hybrid.  Hybrid buses are not quite ready for primetime, but their mass production may not be too far off.  Having buses that produce lower emissions is a significant milestone.  Now, if only something could be done about the smell...

## Traffic and Congestion

My family and I live just twenty-three miles from Chicago's downtown; yet, yesterday it took over an hour and fifteen minutes just to get into the city (though I admit I was foolish enough to leave during rush hour.) When you're moving at a snail's pace on the highway, you have a lot of time to think. Highways were first built during the Eisenhower years and have expanded steadily ever since; but as highways have grown, so, too, has the population, the sprawl, and the congestion. These highways weren't built to handle the high volume of cars that now move through them; we're still driving on designs that were built for a smaller number of cars and less people.

The massive investment needed to get our roads to match our demand may never happen. So we're stuck sitting in our cars for hours; a waste of both fuel and time. What is needed now is a rush hour fee, just like how they do it in London. London charges you a fee for entering the downtown area, which is done to reduce congestion; and it's working. More people are taking public transportation and are driving into the city only when they really need to. This same fee structure could work in our own major cities, too. A rush hour fee is a great way to reduce congestion and move people towards mass transit. It would reduce pollution and lower our dependency on foreign oil.

I'd imagine that instituting a rush hour fee would be a tough sell here in the US; but I for one would welcome the change from sitting bumper to bumper on a daily commute.

**Light Rail**

One of the line items in the government's economic stimulus package provides financial support to Amtrak. The Bush administration didn't provide the necessary funds to help Amtrak improve its operations, but Obama's administration has, yet, the package falls short. I don't believe the political will is there to really back an overhaul of our rail transit systems.

Amtrak and most other commuter lines around the country run on rail lines maintained by freight companies like Union Pacific, CSX, Norfolk Southern, and Canadian National. Freight lines do an excellent job investing in and maintaining their lines. The only issue is that commuter lines like Amtrak must allow freight to take precedence over people. For that reason, we need a national program to build rail services for moving people not only from city to city but within cities, too. A few cities do it very well, like my hometown Chicago. My only criticism is that in Chicago the spokes are not connected; Chicago's commuter lines radiate from the city center as spokes on a wheel. Those spokes need to be connected so people can move easily within the Chicago area without heading into the city center and then back out. London's tube system and the Paris Metro operate on a spoke system as well, but they are all connected.

We need to invest in light rail around the country. High-speed city-to-city connections like Amtrak's Acela Express have excelled in the Northeastern part of the US. Adding express trains to other parts of the country would surely be a massive public works program; but it would cut down on fuel emissions, reduce congestion, provide public work jobs for years to come, and link neighborhoods. A massive rail build-out program would greatly improve our daily lives and should be part of our next green stimulus bill.

## Air Travel

I tend to spend quite a bit of time flying—not an enjoyable experience.  As a frequent flyer, I do get to board the plane first, but still suffer the consequences of poor air quality, delays in air traffic control, security checkpoints, and maintenance problems.  According to reports, air traffic is lower, there are fewer planes in the air, and all seats are full due to capacity cutbacks.  All of these factors lead me to this question: why do we still have non-weather related air traffic delays?

Flying is not pleasant.  The fact that we're stuffed in circular metal tubes shaped like cigars, hurtling through the skies at 560 mph, and breathing recirculated air makes me a bit uneasy.  Flying despite cutbacks hasn't gotten any easier; even the Transportation Security Administration is finding new ways to add to the misery, such as random security checks at the gates.  It's bad enough that we have to take off our shoes, belts, and clothing to get through security.  Why is it that the TSA, airline personnel, and the military are exempt from these requirements?  I mean, who is paying the bill here: them or us?

## Canadian Winters to Test all Winter Cars

If you own a hybrid vehicle and live in a northern climate, you've probably noticed that your fuel efficiency drops during the winter. I've noticed this with the Prius I drive. During warmer months I can get my mileage up into the high 40's, but in the winter it plummets into the 30's, especially on bitterly cold days.

The province of Quebec and a local public utility company plan on testing fifty all-electric vehicles in a partnership with Mitsubishi Motors. The joint $4.4 million project will test the vehicles against the rigors of the Canadian climate and the infrastructure needed to support them. Mitsubishi Motors is providing the province with fifty i-Miev four-door all-electric vehicles; this will be Canada's largest trial of all-electric vehicles. Organizers say the road test is the first to include a car manufacturer, public utility, municipality, and local businesses, which will integrate Mitsubishi i-Mievs into their existing fleets. Mitsubishi says its Innovative Electric Vehicle, or i-Miev, is an all-electric, highway-capable, charge-at-home commuter car. Quebec's provincial government hopes the project will help them identify the infrastructure needed to support all-electric vehicles in homes, workplaces, and public areas. Quebec was Canada's first province to adopt California's strict auto emission standards; the new standards impose increasingly stringent limits on greenhouse gas emissions from cars and light trucks.

US cities facing similar temperature constraints should watch this project carefully. Quebec's experiment could provide the blueprint for colder cities to adopt all-electric vehicles despite the weather extremes they face.[145]

## Audi Green Police Bust

If you happened to be one of the 150 million people who watch the Super Bowl, you may remember having caught an Audi Green Police commercial. The commercial garnered plenty of media attention regarding how Audi is appealing to those consumers wanting to do something green, just not at the level of an Al Gore messianic convert. Audi is touting their new "clean diesel" A3, which coincidentally won 2010's Green Car of the Year, beating the Toyota Prius. I have to say I'm a little shocked that a diesel won Green Car of the Year. Though the A3 gets forty-two miles to the gallon and the diesel is twenty percent cleaner in emissions than previous models, it's still a fume-emitting car. Is it OK for "clean diesels" to be accepted as green cars?

We need to lower and eliminate our emissions. While the new green diesel Audi A3 is an improvement over previous models, it still falls short of what a hybrid can do. Diesels still operate under old principles; they don't turn off at a stoplight or switch to electric mode to burn off excess battery storage. If Audi could create a hybrid diesel car, it would be a great step forward.

The interesting thing about this A3 is that it's not the best car Audi can produce. If you buy the A3 in the US, you can only get an automatic front-wheel drive clean diesel; you can't buy a manual all-wheel drive clean diesel. If Audi had made a manual A3 available in the States, the car's mileage would have gone up to fifty mpg.

Overall, Audi falls short. Trying to make you believe it's OK to be somewhat green is sending the wrong message. Where's the real Green Police when you need them?

**Auto Show Concept Cars Thrill**

The best thing about getting into a car show on media credentials is the quick access to media representatives and the lack of crowds. I began my search by looking for the next generation's green concept vehicles and found a few surprising standouts.

The first kudos goes to Hyundai for their Blue Will concept four-door sedan. Though small, the sedan boasts a list of innovative features: a panoramic glass roof with solar cells for recharging batteries, a thermal generator that converts hot exhaust gases into electricity, lithium polymer batteries, touch-screen controls, and Blue Will, which serves as a test bed of new ideas ranging from roof-mounted solar cells to drive-by-wire steering. In the future, Blue Will could see the road as a hybrid-only model similar to the Toyota Prius. The concept sedan promises a fuel economy rating of more than a hundred miles per gallon and an electric-only driving distance of up to forty miles on a single charge.

Across the aisle from Hyundai was a big sign for Cadillac. Yes, Cadillac; but not your father's Cadillac. Caddy showed off a hybrid concept that was sleek and aggressive in stance. The car had an all-white leather interior and a modern retro look. It was not a large body vehicle, but it seemed very nimble. Cadillac's media personnel indicated that two other current models would soon be offered in hybrid versions. The one hot testosterone vehicle came from Fisker Automotive: an electric hybrid model that goes from 0-60 in seven seconds and looks like something from the movie *Speed Racer*.

When walking through the convention center, I found myself wondering why there were so few alternative-powered cars. Some of the models looked great and I hope we can see them in a hybrid version; but none of them are forthcoming. That is one of the few frustrations of covering a show like this—you just want to see more green options.

## Converting Vehicles

CalCars (*CalCars.org*) is a non-profit startup organization that promotes plug-in hybrids, which they call PHEVs. PHEVs are like regular hybrid vehicles but can be plugged into an outlet and recharged (CalCars even converted a Toyota Prius into a plug-in). Their PHEV Prius gets 100+ miles to the gallon and runs on an electric battery for local driving; but it has the safety and peace of mind of an added on-board gas tank.

CalCars' studies show that the majority of Americans drive less than forty miles a day. The plug-in Prius thus allows drivers to run all of their errands solely on battery power; and the plug-in version needs no dedicated charging stations. It just needs a regular home electrical outlet to charge the car's batteries.

CalCars is promoting their technology to auto manufacturers and large fleet operators. The organization takes the position that even vehicles without hybrid technologies can be converted into all-electric plug-in versions; they insist that even large pickup trucks and truck fleets can take advantage of this technology. The Postal Service is actually asking Congress for the funds to start converting their delivery trucks to PHEVs, estimating that the cost savings in fuel over a five-year period would pay for the conversions. CalCars.org has proposed to auto manufacturers to convert current vehicles and build this feature into new ones. Unfortunately, the current cost of a PHEV runs around $20,000—too high a sum to warrant this type of investment for the average car owner. CalCars emphasizes that if people hung onto their vehicles longer and if gasoline prices were to climb to five dollars a gallon, this type of conversion could pay for itself over time. The problem CalCars faces is the reluctance of auto manufacturers to promote longer ownership of the vehicles they build; but, the PHEV conversion process could ignite an entirely new cottage industry of mechanics that could offer this service once the costs become more reasonable.

## Sustainability at Sixty MPH

Out of the deep forests of Oregon comes one of the most innovative automotive designs in sustainability: a car that can reach a top speed of sixty mph and requires no fossil fuels—or any type of fuel source for that matter. It is not an electric car, a hybrid, or hydrogen vehicle. It's the Human Car, the FM-4 v1.0 (Fully Manual 4 Passenger) to be exact, and the brainchild of engineer Charles Greenwood. He has been working on this design since 1968, when he designed his first three-wheeled car powered by rowing.

The FM-4 v1.0 requires four people to propel it to its top speed: two passengers sit in the front and two in the rear sitting back to back. The passengers power the car by pushing and pulling t-bars. The front passengers sit in concave pads and swivel from side to side; they act as the steering mechanism, similar to the way a person maneuvers when skiing or snowboarding. There is an electric-assist motor to help the car when it's on steep hills; the electricity generated when the car is coasting is stored in a battery. The FM-4 v1.0, while lightweight, is strong and very safe in collision situations, according to Greenwood. The weight of it could reduce noise pollution, eliminate $CO_2$ emissions, and save billions of dollars in road wear and repair.

The cost of the FM-4 v1.0 runs around $6,500; plans are in the works to obtain $10 million in financing to build a plant to mass-produce them. Despite being an open-air vehicle, the cars will still come with GPS and Bluetooth capabilities; there is a version with a solar roof for cover and a model made from recycled plastics. One of the best things about the car is that it can be hooked up to generate electricity for your home once it's parked in the garage. Four people working in tandem can generate 1,200 watts of electricity within three minutes of pushing and pulling in a stationary position, which is enough power to run your TV for thirty minutes.[146]

## Fleet Buyers Can Jump-Start Electrics

The Electrification Coalition, a group made up of CEOs from the automotive and electric power industries, released a roadmap for fleet electrification, arguing that the use of corporate fleets can help make plug-in vehicles more commonplace. Members at the November, 2010 press conference in Washington, DC, released the document in an effort to influence policy.

According to coalition member Frederick Smith, Chairman, President, and CEO of FedEx, electrifying transportation may be the best way to reduce our oil consumption. That, along with fuel efficiency mandates, means that the US could decrease the oil consumption of light-duty vehicles from around ten million barrels of oil today to four million barrels within twenty-five years, Smith said. By 2035, consumption in the US is expected to hit approximately sixteen million barrels of oil per day, which is one factor leading to expected increases in oil prices.

The fleet roadmap analysis finds that EVs, short for electric vehicles, are cost-competitive in many fleet applications today. According to the report, EVs will become the most cost-effective option as battery prices go down within the next five to eight years; by 2012, standard hybrids driven over 20,000 miles a year may become more economical than internal combustion engines.

In 2009, there were 16.3 million fleet vehicles in operation. Having fleet operators jump-start EV's would go a long way towards helping drive the cost down for the rest of us. The fact that commerce is willing to get behind this is quite heartening. It's time for all of us to join them in this effort.[147]

## Chapter Thirteen: Innovation

"What we do today, right now, will have an accumulated effect on all of our tomorrows."

- Alexandra Stoddard, *philosopher, author, and interior designer*

## Body Heat

Where can you find cheap renewable energy sources, you ask? Would you believe that one such source actually comes from your own body heat? Sweden, one of the world's coldest climates, has found an innovative way to heat the main train station in its capitol, Stockholm. Over 250,000 commuters crowd into Stockholm's Central Station each day. Body heat generated by this mass of individuals previously escaped into the air; but now the heat is being captured and utilized from one building to another. It isn't the first time body heat has been used for heating a building. The Mall of America in Minneapolis uses body heat to regulate the mall's temperature; but what sets Stockholm's train station apart is that it is the first building in which body heat actually moves from one building to another.

The new heating system in Stockholm began operating in April of 2010 and works in the following manner: heat generated by the commuters is first captured by the station's ventilation system and used to warm water in underground tanks. The water is then pumped through pipes to another building that stands a hundred yards from the train station, where it is used in the building's main heating system. The commuters' body heat is estimated to provide up to thirty percent of the building's heat.

The company that owns the train station also owns the building being heated by the station. According to the engineers who worked on the project, the movement of body heat for heating purposes works because the buildings are close to each other (low temperature heat can only be moved if buildings are close together). The system also works in Stockholm's Central Station because of the volume of people moving through it. One of the great things about the system is that it can be applied to buildings in other parts of the world as long as there are large numbers of people moving through them on a daily basis. Think what this could do for major cities in the US...[148]

**New Electric Scooters**

Electric scooters are all the rage in suburban neighborhoods; kids have been tooling around on inexpensive Razor electric scooters for years. Now Honda, Japan's number-two automaker and the number one motorcycle manufacturer in the world, plans to release a zero-emission electric scooter. Named the EV-neo, Honda's scooter is equivalent to a 50cc gas-powered bike. It runs on a lithium-ion battery and can be fully charged with a normal socket in four hours or recharged to about eighty percent capacity within twenty minutes using a rapid charger. The range of the EV-neo scooter is currently nineteen miles.

The super quiet EV-neo is initially being marketed to commercial users; commercial enterprises that depend on local delivery service, like pizza restaurants and newspapers, are the scooter's main target audience. Honda first plans on leasing the EV-neo to companies, and eventually to consumers, in Japan and China. According to a Honda spokesperson, the company isn't currently looking to sell its scooters outside of those two markets.

Honda hopes to set a price for the EV-neo lower than a regular scooter over a three year period, including gas prices (a range that would be about 600,000-800,000 yen [$6,000-$8,000] per scooter). Rival Yamaha Motor Co., the world's second-biggest motorcycle maker, is also scheduled to begin selling a new battery-run scooter in Japan in late-2010; they also plan to launch in Taiwan and Europe as well. Yamaha had to discontinue a range of electric scooters sold in Japan between 2002 and 2006 after their lithium-ion batteries were recalled.[149]

## Green Roofs

One of the most famous green roofs in the country can be found at Chicago's City Hall. Mayor Richard Daley had the roof converted several years ago—and it's been successful at insulating the building during the winter and summer months. [150]

Not many developers have made the transition to providing green roofs for new projects. Given the current economic downturn, some may not even consider them for future projects; but it would be a step in the right direction, though, if new building projects began development using green roofs and had LEEDS certification.

I recently discovered something growing in my garden that happens to be a key component in green roof technologies: sedum. Sedum is a ground cover plant that proliferates like a weed. It's a hardy plant that can tolerate extreme temperature fluctuations and drought-like conditions. The plant's root structure compacts into a tightly woven mat and can survive being walked on. If by chance any of you are familiar with garden railways, you may know that sedum is one of the chief plants used as ground covers. It looks like a miniature forest and has amazing endurance (I'm always trying to figure out how to keep it from taking over the yard!). Now I know where I can really use it—on the roof.

Green roofs could go a long way towards reducing our carbon footprint. They would absorb the sun and keep global warming in check. Green roofs have also been found to last as long as conventional roofs and can be easily recharged at any time; besides, having a green roof looks a lot prettier than an asphalt one.[151]

## Orange Peel Tires

Tire manufacturer, Yokohama, has produced a tire made from eighty percent sustainable ingredients derived from orange oil and rubber. The tire model is called Super E Spec, which received the Popular Mechanics Editor's Choice Award in 2008. Yokohama is primarily marketing the tire to hybrid vehicles (why wouldn't they market it to other types of vehicles, too?). The Toyota Prius Hybrid is scheduled to be the first car to utilize the tire.

According to Yokohama, there is a twenty percent reduction in rolling resistance with orange peel oil tires, which should result in savings at the gas pump. How much in savings at the gas pump is not mentioned. Current tires are the bane of landfills. They aren't easily recycled due to the cost of separating rubber from petroleum. Additional hazards to tire recycling involve the burn factor: tires burn for long periods of time and release toxins into the air. Orange oil, on the other hand, is derived from orange peels and easily biodegrades.

Yokohama has not yet released an environmental impact statement on the disposal of these tires. Once released, it will be interesting to see the environmental impact statement of this kind of tire and just how sustainable it really is; but, if it lowers our dependency on petroleum, all the better.[152]

## 7,000 Gallons

The biggest water users in the world can be found in the US, Canada, Japan, and Mexico. The average person living in Haiti uses 1,500 gallons of water a year for all of his or her needs; the average American, on the other hand, uses 7,000 gallons of water per year to flush the toilet. 7,000 gallons just to flush the toilet! That's a pretty staggering number, even when you take into account the low flow toilets being installed today. There are alternatives, though, to all of that flushing. One such alternative is composting toilets, which use no water; but how do they work—and would you want one in your home?

Composting toilets dehydrate and compost human waste; by doing so, they actually create a usable byproduct that can be used in flowerbeds or as a soil additive. This all may sound gross, but according to the manufacturers, the toilets are extremely sanitary and odorless due to a built-in airflow system. Still, there is a little maintenance required: you need to clean out the composting tray at the bottom of the toilet every few weeks, since it contains a humus-rich nutrient soil.

The composting toilets come in different sizes and can fit easily in your bathroom. If you can get over the grossness-factor, it's actually a great way to green your home.[153]

**New "Tree" Carbon Catcher**

A new synthetic device is being developed that can capture carbon dioxide emissions 1,000 times faster than a real tree. Klaus Lackner, a professor of geophysics at Columbia University, has been working on the project since 1998, according to a CNN report. The device itself operates like a sponge, drawing in carbon dioxide emissions, cleaning them, and then releasing the gas back into the atmosphere. Since coal-fired emissions have been reduced, the main culprits that the device is aimed at are car and airplane emissions. Lackner stated that the device is extremely flexible and can be placed anywhere; the best part is that it would operate all day, everyday, unlike trees, which need sunlight, seasons, and water. There have been no cost estimates on producing the new device, or a timetable for when it will become commercially available. In the meantime, though, let's keep planting those trees; they are sure to be more aesthetically pleasing to the eye.[154]

**US Army to Test Wind Turbine**

The US Army plans to test a mobile wind turbine to power communications equipment; the trailer-mounted wind turbine is expected to be tested at an army facility in Maryland. WindTamer, a Rochester, NY-based company, developed the one kilowatt wind turbine for the military. The design is unique in its ability to channel wind through a shroud to increase the efficiency of the turbine; WindTamer's turbines collect wind through a housing that fits around fan blades. According to the company, when wind passes through the "shroud", it creates a pulling effect to draw more air and generate more electricity. Since most small turbines need a fast wind before they can begin operating, the design makes placing wind turbines in less windy places more practical.

One of the key features of WindTamer's turbine is its relatively low profile. The lower height of the turbine could bypass local ordinances covering building height restrictions and make it for desirable for homeowners.

WindTamer is just one of a handful of companies exploring alternatives to the traditional "open rotor", three-blade wind turbine. Other companies that have developed similar style turbines include FloDesign Wind Turbines, WindTronics, and OptWind.[155]

**Pimp Your Rickshaw**

The Japanese have developed a new electric concept for one of their oldest methods of transportation: the rickshaw. Two companies from Osaka, Japan, have created a modern electric lithium-ion battery-powered rickshaw made from metal, bamboo, and paper. The three-wheel vehicle is called a Meguru, the Japanese word meaning "to move". Its steel frame is painted with vermillion lacquer and its flooring is bamboo; the folding doors are made from paper and resemble a Japanese hand fan. The cool thing about the vehicle is that the interior lights glow through the paper doors at night, giving it a lantern-like appearance. It is one pimped out rickshaw! I think it will end up finding a new global inner city tourist audience due to its pollution-free capabilities and cute design.

The Meguru's lithium-ion battery takes two hours to charge on a household power supply and can travel twenty-five miles on a single charge. It has a top speed of about twenty-five mph and is registered as a road vehicle in Japan. While a novel idea, replacing foot power with an all-electric version does add an additional burden to the current power grid: most of those plants are powered by coal.[156]

## Amtrak's New "Cattle" Train

Amtrak has begun a pilot program to operate a train line that runs on biodiesel fuel. The biodiesel is being manufactured out of Texas using beef byproducts. The Amtrak train, called the Heartland Flyer, is running on the fuel and operates daily from Oklahoma City to Fort Worth, TX. According to Amtrak, the biodiesel, which mixes eighty percent diesel with twenty percent biofuel, cuts both hydrocarbon and carbon monoxide emissions by ten percent. The company also stated that, compared to standard diesel fuels, biodiesel reduces particulates by fifteen percent and sulphates by twenty percent.

Amtrak is running the train service as part of a twelve-month experiment on the use of alternative fuel sources. During this time, the company will collect data on emissions and on the impact of the fuel on mechanical parts. Although technically the fuel mix can run in unmodified trains, the locomotive was fitted with new engine assemblies so that detailed measurements could be taken to establish the effect of the fuel on the engine. The impact of biofuel blends on engines can vary dramatically; some biofuels lead to increased wear and tear, while others tend to burn cleaner and lead to improved engine performance and durability.

The biodiesel trial is the latest in a series of environmental initiatives from Amtrak designed to highlight the operator's position as a green alternative to domestic air travel. The company has switched from using low-sulphur fuel to ultra-low sulphur fuel to tackle air pollution and has installed recycling receptacles throughout its trains and stations. They have also stepped up efforts to reduce idling times and introduce regenerative braking systems for their electric trains, similar to those found in hybrid cars.[157]

**Terra Cycle Turning Waste in Product**

Terra Cycle is a new innovative new company that is taking waste and turning it into new consumer products. The New Jersey-based company, founded in 2001 by a nineteen-year-old Princeton graduate, has become one of the most successful companies at turning waste into consumer products. They have struck up partnerships with companies like Capri Sun and Frito Lay to convert their packaging waste into reusable products; Terra Cycle recycles finished Capri Sun juice boxes and Frito Lay chip bags back into colorful backpacks and messenger bags. The backpacks and messenger bags literally sport the spent packaging logos, enabling both brands to grab some extra advertising support. Terra Cycle will pay you for your waste, spending two cents per chip bag and juice box. Their website boasts that over eight million people have contributed waste to their recycling efforts. The company has also donated over $500,000 in funding to worthwhile causes.

Terra Cycle products are sold through traditional retailers like Wal-Mart and Home Depot, as well as through their online store. They are seeking companies, and people, to sign up and partner with them in producing new products from recycled waste.

New Jersey is called the Garden State. Trust me on this; get past Newark and you'll find it a beautiful state. Now Terra Cycle adds one more accolade to the state motto.[158]

## Human Waste: The Next Premium Biogas

The electrification of automobiles has been around since the end of the late 1800s, with bio-diesel efforts beginning in the 1970's. Nowadays, we have natural gas, gas-electric hybrid cars, and soon the first true electric cars since the invention of the electric golf cart. The race to find a nearly inexhaustible supply of alternative fuel for cars has lead scientists down some interesting paths of discovery. One of their newest discoveries is of an even better means of creating fuel to power cars: using human waste.

Engineers from Geneco, a large biotechnology company operating in 120 countries, have unveiled a modified Volkswagen Beetle that runs on compressed methane gas extracted from human waste; the waste itself comes from a sewage works processor. Geneco's biogas is made by putting human waste in a decomposition chamber through which oxygen-starved bacteria can break it down to produce methane. The methane is then harvested and placed in tanks inside the trunk of the car where it's used to power the slightly modified engine. One of the neat things about it is that if you run out of biogas, your car can still run on regular gasoline.

According to Geneco, seventy homes can generate enough gas to run a car as far as 10,000 miles. They have also said that the Volkswagen is carbon neutral; it produces less carbon dioxide than a normal car would in an average year (3 tons vs. 3.5 tons). Since we are going to produce that methane anyway, why not use it in a carbon neutral manner? Other good (and reassuring!) news is that the methane doesn't smell.

Biogas isn't really a new invention, as similar efforts have included using cow dung; yet, this is the first time that the use of human waste has been tried. If human biogas ever becomes a practical application, I'd venture to guess that eating spicier foods would help accelerate our efforts to become carbon neutral.[159]

## Shreds, Bends, and Wipes

What shreds, blends, dries, and wipes your behind? It's the next inventive machine to come out of Japan (no, it's not an extra in the latest Godzilla movie). The machine is called the White Goat, and it's the next best thing to an office shredder. The White Goat, made by the Oriental Co., was first featured at a Japanese ecological trade show in 2009. It works by taking everyday office paper, processing it into a pulp, sending it through a drying process, and then rolling it into toilet paper. Now your office memos and sales pitches can be processed into white toilet paper.

Each roll of toilet paper is made from forty sheets of paper; the entire process takes about thirty minutes per roll. Oriental Co. has filed for a patent on their machine and claims that by recycling office paper we can save sixty trees annually; all you have to do as a consumer is provide water and electricity. The White Goat is fairly large and would need at least a ten by ten square foot area. It also comes with a price tag of $100,600.00.

In reality, even if the machine ran twenty-four hours a day, it still could only produce forty-eight rolls of toilet paper; and it would take years to recoup the cost. You have to give the Japanese credit, though, in creating a machine that not many of us would have even thought of. It's a great approach to finding a new way to recycle all the office paper we still shuffle around.[160]

## Vertical Farming

By 2050, there may be another three billion people added to our current global population. We will need a landmass the size of Brazil to feed them. Besides finding a way to curb this population surge, we will need to rethink how we feed ourselves. Dr. Dickson Despommier, Ph.D at Columbia University, is one such person who is rethinking things. He has developed the concept of Vertical Farming, which takes its cue from hothouse farming methods. It is a potential way of growing food on a 24/7, 365 days per year basis.

The crops grown in vertical farms are taken from a list of crops deemed by NASA to be viable in hydroponic growing conditions. The farms themselves are controlled environments that can generate their own power and recycle the water used in growing food crops. In a controlled environment, there will be no impact from pests or from changing weather conditions. The result is a secure food production environment.

Vertical Farming is basically an urban built oasis—kind of like a tall skyscraper that can feed 50,000 people. It takes up less land to feed a large number of people and reduces transportation costs for getting food to local markets. Benefits could also include taking obsolete buildings and converting them to vertical farms.

In certain parts of the world, vertical farms could allow us to restore food-growing areas to forests; new forests would help solve our global warming problem by reducing $CO_2$ levels around the globe. Vertical farms would also help world soils replenish themselves the way the US solved the dust bowl problems in the 1930's.

To get a great education on vertical farms, visit *www.verticalfarm.com*. They are our future.[161]

## Turning Electronic Waste into a Bike Path

Electronic waste, better known as e-waste, is becoming a landfill problem for communities around the world. The majority of old inkjet printer cartridges, for instance, are ending up in landfills. On the bright side, though, there has been an increase in efforts to recycle electronic goods and find new uses for e-waste.

Inkjet printer manufacturers have been at the forefront of recycling print cartridges and finding ways to "green" their existing cartridges. The Australian National Park Service may have even found the perfect solution to reusing spent inkjet and laser cartridges. The Park Service, in a partnership with Repeat Plastics Australia, opened a new bike path made entirely out of discarded printer cartridges. The 10.6-mile bike path stretches between Alice Springs and Simpsons Gap in the northwest territory of Australia and sees over 120,000 visitors every year. According to Parks and Wildlife Minister Karl Hampton, the bridge echoes the Australian government's commitment to sustainable development: "Saving landfill, trees, and ensuring a longer life with less maintenance." The bike path has a cost effective component to it as well. The entire path, complete with a viewing platform, only cost the park service $330,000 to complete.

We've seen recycled plastic made into park benches, playgrounds, and parking bumpers; but this has to be the first bike path made from recycled plastic. Imagine if we could use recycled plastic to manufacture future road surfaces and even cut down on the amount of asphalt needed for resurfacing. It could provide us with a longer road lifespan than current asphalt.[162]

## The Age of Sail is Back

Commercial ocean shipping carries up to 98.2 percent of the world's goods, 98 percent of which is driven by diesel engines. Ocean bound freighters are contributing three percent of carbon emissions to the atmosphere. If all commercial ships were a country, they would rank number seven in total carbon emissions; but there may be help on the way to cut these emissions in half. SkySails, a German company based in Hamburg, has developed a 21$^{st}$ century solution to an ancient technology: sail power. Although harvesting wind power to cut down on carbon emissions is not a novel idea, SkySails' concept involves using a large towing kite to help propel the ship and cut down on diesel fuel usage. The age of sail may again be upon us.

The German company claims that almost every merchant and passenger vessel can be equipped or retrofitted with the SkySails system. The towing kite is filled with compressed air to obtain optimal-shaped aerofoil profiles and can deliver up to 5,000 square meters of sail; a computer autopilot operates it for the best possible energy utilization. The operating altitude of the scales can run up to 500 meters; the speed of the wind increases with height, even at heights with clam winds. SkySails believes sufficient wind energy is available to enable ship owners to stay on schedule and halve their fuel costs.

SkySails received government sponsored funding for the project. The real question is whether the shipping industry is willing to spend the money to retrofit their ships. Shipping took a hit during the last recession and has cut back on the number of ships they're deploying; consequently, it's taking longer for merchandise to cross the oceans. Someone has to hurry up and do the cost benefit analysis on this one in order to convince ship owners to make the change. It would be very neat to see large sailing ships traversing across the oceans again.[163]

## These Boots are Made for Charging

Orange, UK, a cellular provider in the United Kingdom, has unveiled one of the coolest devices we've seen in a longtime. At a festival in the UK, Orange provided the attendees with a novel way to keep their iPods and mobile phones fully charged during the event. Utilizing a Wellies rain boot, coupled with a thermoelectric charger, the boots would charge their electronic device as they sloshed around the festival.

The Orange Power Wellies, created in collaboration with renewable energy experts GotWind, use a unique "power generating sole" that converts heat from your feet into an electrical current. This 'welectricity' can then be used to recharge your mobile phone. Twelve hours of stomping through the festival in your Orange Power Wellies gives you enough power to charge your mobile device for one hour. To increase the length of time you can charge your phone simply hot step it to the dance tent, since the hotter your feet get, the more energy you produce. After a full day of festival frolicking, attendees could plug their phones into the power output at the top of the welly and use the energy generated throughout the day to charge their phones.

The power collected in the 'power generating sole' is collected via a process known as the 'Seebeck' effect. Inside the power generating sole there are thermoelectric modules constructed of pairs of p-type and n-type semiconductor materials forming a thermocouple. These thermocouples are connected electrically, forming an array of multiple thermocouples (thermopile). They are then sandwiched between two thin ceramic wafers. When the heat from the foot is applied to the topside of the ceramic wafer and cold is applied to the opposite side, electricity is generated. It's a great sustainable way of powering electronic devices without tapping into the power grid. The only problem is that Orange, UK, does not seem to have any plans to market the product to consumers in the immediate future.[164]

## 2010 Solar Decathlon

2010's Solar Home Decathlon took place in Madrid, Spain, and marked the first time the solar exhibition took place in Europe. Seventeen teams from around the world entered eco-friendly home designs into the competition; all entrants were required to build their entry for consideration to win the Solar Decathlon. Two of the seventeen entrants were from the United States: the University of Florida and Virginia Polytechnic Institute & State University.

The rules for entrance into the Solar Home Decathlon required that all solar homes be of prefabricated construction. The teams were then graded on their sustainability, innovative architecture and engineering, and their ability to minimize energy use. The solar designs presented were cutting edge and stood out from traditional home designs. They represent the future for a sustainable carbon dioxide-free existence in home building and design.

These are exciting times for the green movement. With new solar home design options growing for consumers and the release of future hybrid and all-electric vehicles, we will be able to make choices in sustainability that have long been few and far between. The advent of new eco-friendly choices for consumers only hastens the demise of a carbon fuel-based society. That in itself makes it worth taking a look at these designs.[165]

1       COUNTRY SCORES. *Environmental Performance Index 2010*. 2010. http://epi.yale.edu/Countries  Retrieved 11/4/10

2       EPI Scores (By Rank). 2010 ENVIRONMENTAL PERFORMANCE INDEX. http://www.ciesin.columbia.edu/documents/EPI_2010_report.pdf retrieved 11/4/10

3       [3] Feeding Birds Could create a New Species. Brandon Keim. *Wired*. December 3, 2009. http://www.wired.com/wiredscience/2009/12/bird-feeding-evolution/ retrieved 11/4/10

4       Climate Changes Confuses Migrating Birds. Jasper Copping. *Telegraph.co.uk*. Published 3/16/08. http://www.wired.com/wiredscience/2009/12/bird-feeding-evolution/ retrieved 11/4/10

5       Climate Change to Trigger Wave of Refugees-UNHCR. Newton Sibanda. *The Daily IIJ*. October 15th, 2010. http://inwent-iij-lab.org/Weblog/2010/10/15/climate-change-to-trigger-wave-of-refugees-unhcr/#more-4438 retrieved 11/4/10

6       Fleeing Drought in the Horn of Africa. Edmund Sanders. *Los Angeles Times*. October 25, 2009. http://articles.latimes.com/2009/oct/25/world/fg-climate-refugees25 retrieved 11/4/10

7       Adding Sulfur to the Atmosphere Proposed to Ease Global Warming. *Earth Observatory*. July 27, 2006. http://earthobservatory.nasa.gov/Newsroom/view.php?id=30475 retrieved 11/5/10

8       Artificial Volcanoes – Reflective Particles in the Stratosphere. Geopiracy- The Case Against Geoengineering, p.25. *Etc Group*. 10/18/2010. http://www.etcgroup.org/en/node/5217 retrieved 11/5/10

9       Deserts Shrinking Because of Global Warming? Ned Haluzan. *Ecological Problems Blog*.
July 17, 2009. http://ecological-problems.blogspot.com/2009/07/deserts-shrinking-because-of-global.html retrieved 11/5/10

10      38% of World's Land in Danger of Turning into Desert. Brian Merchant. *Treehugger*. February 2, 2010. http://www.treehugger.com/files/2010/02/38-worlds-land-danger-turning-desert.php retrieved 11/5/10

11      The No Impact Experiment. Colin Beavan. *No Impact Man*. February 21, 2007. http://noimpactman.typepad.com/blog/2007/02/the_no_impact_e.html Retrieved 11/5/10

12	Researchers Find Way to Cut Methane Gas in Cattle. Bev Betkowski. *University of Alberta Express News.* May 8, 2009 http://www.archives.expressnews.ualberta.ca/article/2009/05/10171.html retrieved 11/5/10

13	History of Earth Day. *Earthday Network.* 2009. http://www.earthday.net/node/77 retrieved 11/5/10

14	Earth Day and EPA History. *United States Environmental Protection Agency.* July 30, 2010. http://www.epa.gov/EarthDay/history.htm retrieved 11/6/10

15	Eastern US Forests Resume Decline. *American Institute of Biological Sciences.* April, 2010. http://www.aibs.org/bioscience-press-releases/100407_eastern_us_forests_resume_decline.html retrieved 11/6/10

16	Earth Has Entered The "Anthropocene Epoch," Geologists Say. *Science 2.0.* March 26, 2010. http://www.science20.com/news_articles/earth_has_entered_anthropocene_epoch_geologists_say retrieved 11/6/10

17	Dawn of the Anthropocene Epoch? Earth Has Entered New Age of Geological Time, Experts Say. *Science Daily.* March 26, 2010. http://www.sciencedaily.com/releases/2010/03/100326101117.htm retrieved 11/6/10

18	Are Earthquakes Really on the Increase? *U.S. Geological Survey* http://earthquake.usgs.gov/learn/topics/increase_in_earthquakes.php retrieved 11/8/10

19	Is Recent Earthquake Activity Unusual? Scientists Say No. *U.S. Geological Survey.* April 4, 2010. http://www.usgs.gov/newsroom/article.asp?ID=2439 retrieved 11/8/10

20	Gulf Oil Spill Exceeds BP's 'Worst-Case Scenario,' Drilling Supporters On Defensive. Marcus Baram. *The Huffington Post.* April 29, 2010. http://www.huffingtonpost.com/2010/04/29/gulf-oil-spill-exceeds-bp_n_556798.html retrieved 11/21/10 retrieved 11/8/10

21	Avatar Sands. *Dirty Oil Sands.* http://dirtyoilsands.org/files/AVATARSANDS_Variety_Final_PRINT.pdf retrieved 11/6/10

22	Avatar(Sands) Has our Vote. *Environmental Defence.* March 04, 2010. http://environmentaldefence.ca/blog/avatarsands-has-our-vote retrieved 11/6/10

23      The Carbon Footprint of Email Spam Report. *Sustaincommworld*. 2009 www.sustaincommworld.com/.../Email_Carbon_Footprint_Report.pdf retrieved 11/6/10

24      Greening the Internet: How much CO2 does this article produce? Lara Farrar. CNN. July 10, 2009 http://articles.cnn.com/2009-07-10/tech/green.internet.CO2_1_greenhouse-gas-emissions-carbon-footprint?_s=PM:TECH retrieved 11/6/10

25      Haircare ingredients could hold key to reducing CO2 emissions. Jeff Salton. *Gizmag*. March 24, 2010. http://www.gizmag.com/reduce-co2-emissions-coal-power-stations/14614/ retrieved 11/6/10

26      New CO2 'Scrubber' from Ingredient in Hair Conditioners. *Science Daily*. March 25, 2010. http://www.sciencedaily.com/releases/2010/03/100324141953.htm retrieved 11/6/10

27      New USGS Study Documents Rapid Disappearance of Antarctica's Ice Shelves. *U.S. Geological Survey*. April 3, 2009 http://www.usgs.gov/newsroom/article.asp?ID=2186&from=rss_home retrieved 11/8/10

28      NOAA: U.S. Had Eighth Warmest June on Record, Above-Normal Precipitation. National Oceanic and Atmosphere Administration. July 12, 2010. *High Plains Regional Climate Center*. http://www.hprcc.unl.edu/articles/index.php?id=183 retrieved 11/8/10

29      ''We don't ground all planes'' analogy crashes and burns. Mark Moseley. *The Lens*. July 14,2010. http://thelensnola.org/2010/07/14/plane-analogy/ retrieved 11/8/10

30      Oil rigs: Cities at sea. *The Week*. May 14, 2010. http://theweek.com/article/index/202892/oil-rigs-cities-at-sea retrieved 11/8/10

31      Advocates for Electronics Producer Responsibility Speak Out Against NYC e-Waste Lawsuit. Jaymi Heimbuch. *Treehugger*. January 14, 2010. http://www.treehugger.com/files/2010/01/advocates-for-electronics-producer-responsibility-speak-out-against-nyc-e-waste-lawsuit.php retrieved 11/8/10

32      Clean Energy and Clean Shoes in San Francisco. *US EPA*. March 31, 2010. http://www.epa.gov/wastes/inforesources/news/2006news/06-dog.htm retrieved 11/8/10

33	What is the Great Pacific Ocean Garbage Patch? Russell McLendon. *Mother Nature Network*. February 24, 2010. http://www.mnn.com/earth-matters/translating-uncle-sam/stories/what-is-the-great-pacific-ocean-garbage-patch# retrieved 11/8/10

34	De-mystifying the "Great Pacific Garbage Patch" *NOAA'S National Ocean Service*. Revised September 18, 2010. http://marinedebris.noaa.gov/info/patch.html retrieved 11/8/10

35	Beached Grey Whale In Seattle Had Shocking Amount Of Garbage In Its Stomach. *The Huffington Post*. April 19, 2010. http://www.huffingtonpost.com/2010/04/20/beached-grey-whale-in-sea_n_544130.html retrieved 11/8/10

36	International Coastal Cleanup 2010 Report: Trash Travels. *Ocean Conservancy*. http://www.oceanconservancy.org/site/PageServer?pagename=program_marinedebris_ICCreport retrieved 11/9/10

37	EcoATM pays you for used gadgets. Leslie Katz. *CNet News*. October 3, 2009. http://news.cnet.com/8301-17938_105-10366816-1.html retrieved 11/9/10

38	Endangered Species: Nine Animals That May Not Survive The Next Decade. Updated: 03-18-10. *The Huffington Post*. http://www.huffingtonpost.com/2009/12/27/endangered-species-nine-a_n_387202.html retrieved 11/9/10

39	Honey Bee Colony Collapse Disorder. Renee Johnson. *Congressional Research Service*. January 7, 2010 http://www.fas.org/sgp/crs/misc/RL33938.pdf retrieved 11/9/10

40	Sex life may hold key to honeybee survival. *University Of Leeds*. September 14, 2009. http://www.leeds.ac.uk/news/article/130/sex_life_may_hold_key_to_honeybee_survival retrieved 11/9/10

41	42 tons of poison to be dumped on island to eradicate rats. Bryan Nelson. *Mother Nature Network*. January 13, 2010. http://www.mnn.com/earth-matters/wilderness-resources/stories/42-tons-of-poison-to-be-dumped-on-island-to-eradicate retrieved 11/9/10

42	Endangered insects to be reintroduced on RSPB Scotland reserves. *The Royal Society for the Protection of Birds*. January 15, 2010. http://www.rspb.org.uk/news/details.aspx?id=tcm:9-239175 retrieved 11/9/10

43	Zoo to bring dead animals back to life, 'Jurassic Park'-style. Tom Chivers. July 1, 2010. *The Telegraph*. http://www.telegraph.co.uk/science/dinosaurs/7865765/Zoo-to-bring-dead-animals-back-to-life-Jurassic-Park-style.html retrieved 11/9/10

44	Summary Statistics. IUCN Red List Version 2010.4. http://www.iucnredlist.org/documents/summarystatistics/2010_4RL_Stats_Table_1.pdf retrieved 11/9/10

45	Extinction crisis continues apace. *IUCN Species Survival Commission*. November 3, 2009. http://www.iucn.org/about/work/programmes/species/red_list/?4143/Extinction-crisis-continues-apace retrieved 11/9/10

46	Assistant Secretary Strickland Announces Continued U.S. Support for Bluefin Tuna Protection. *U.S. CITES*. March 3, 2010. http://www.uscites.gov/position/assistant-secretary-strickland-announces-continued-us-support-bluefin-tuna-protection retrieved 11/9/10

47	Atlantic Bluefin Tuna Proposal Not Adopted After Intense Debate. *U.S. CITES*. http://www.uscites.gov/update/atlantic-bluefin-tuna-proposal-not-adopted-after-intense-debate retrieved 11/9/10

48	Scientists Find Evidence That Oil And Dispersant Mix Is Making Its Way Into The Foodchain. Dan Froomkin. July 29, 2010. *The Huffington Post*. http://www.huffingtonpost.com/2010/07/29/scientists-find-evidence_n_664298.html retrieved 11/9/10

49	Acidification of Oceans May Contribute to Global Declines of Shellfish. October 3, 2010. *Science Daily*. http://www.sciencedaily.com/releases/2010/09/100928154754.htm retrieved 11/9/10

50	Gulf Oil Spill Hits Louisiana's Largest Pelican Nesting Area. Matthew Brown. July 14, 2010. *The Huffington Post*. http://www.huffingtonpost.com/2010/07/14/louisiana-oil-spill-pelican-nesting_n_647002.html retrieved 11/9/10

51	Oil Hits La.'s Largest Seabird Nesting Area. *CBS News*. July 15, 2010. http://www.cbsnews.com/stories/2010/07/14/national/main6679576.shtml retrieved 11/9/10

52	Pollution makes quarter of China water unusable. David Stanway & Miral Fahmy. July 27, 2010. *Environmental News Network*. http://www.enn.com/pollution/article/41584 retrieved 11/13/10

53    The Corrosion of America. Bob Herbert. October 26, 2010. *The New York Times*. http://www.nytimes.com/2010/10/26/opinion/26herbert.html?_r=2&scp=1&sq=corrosion%20herbert&st=cse retrieved 11/12/10

54    "Vast Array" of Drugs in Your Drinking Water. *Shirley's Wellness Café*. August 22, 2010. http://www.shirleys-wellness-cafe.com/water-quality.htm retrieved 11/12/10

55    Desalination Raises Environmental, Cost Concerns. Ben Block. April 29, 2008. *Environmental News Network*. http://www.enn.com/lifestyle/article/35513 retrieved 11/12/10

56    Is Desalination The Answer? Yermi Brenner. August 12, 2010. *The Huffington Post*. http://www.huffingtonpost.com/yermi-brenner/is-desalination-the-answe_b_678968.html retrieved 11/12/10

57    The Interagency Ocean Policy Task Force. Council on Environmental Quality. http://www.whitehouse.gov/administration/eop/ceq/initiatives/oceans retrieved 11/13/10

58    Law Of The Sea. U.N. General Assembly. November 23, 2009. http://www.unga-regular-process.org/images/Documents/law_of_sea_a64_l18.pdf Retrieved 11/13/10

59    Algae: Another way to grow edible oils. Michael Kanellos. Jan 25, 2008. *CNET News*. http://news.cnet.com/Algae-Another-way-to-grow-edible-oils/2100-11395_3-6227572.html?tag=mncol;3n retrieved 11/21/10

60    Peak predictions: mixing water and oil as global resources dwindle. Matthew Wild. *Energy Bulletin*. August 12, 2010. http://www.energybulletin.net/53777 retrieved 11/13/10

61    Our Water Footprint. Simran Sethi. October 25, 2010. *Mother Earth News*. http://www.motherearthnews.com/blogs/blog.aspx?blogid=2147483961 retrieved 11/13/10

62    SODIS- Clean Drinking Water in 6 hours. *SODIS 2010*. http://www.sodis.ch/index_EN retrieved 11/13/10

63    The Water Crisis. Gilbert M. Grosvenor. March 22, 2010. *The Huffington Post* http://www.huffingtonpost.com/gilbert-m-grosvenor/the-water-crisis_b_506648.html retrieved 11/13/10

64    MISSION & PRINCIPLES. *Surfrider Foundation 2010*. http://www.surfrider.org/whoweare2.asp retrieved 11/13/10

65      Australian Dust Storms Feed Life Explosion. Dani Cooper. October 7, 2009. *Discovery News.* http://news.discovery.com/earth/dust-storms-australia.html retrieved 11/13/10

66      Phytoplankton worldwide have been shrinking for 100 years, study shows. Amina Khan. July 29, 2010. *The Los Angeles Times.* http://articles.latimes.com/2010/jul/29/science/la-sci-phytoplankton-20100729 retrieved 11/13/10

67      Press release EnergyMixx AG. May 12, 2009. *Energy Mixx.* http://www.energymixx.com/uploads/media/Pressrelease_EMixx_20090512.PDF retrieved 11/13/10

68      Abengoa Solar IST. Abengoa Solar, Inc. 2008 http://www.industrialsolartechnology.com/solarthermalexperience/projects.html retrieved 11/13/10

69      MIT researchers print solar cell on paper. Martin LaMonica. May 5, 2010. *CNet News.* http://news.cnet.com/8301-11128_3-20004170-54.html retrieved 11/13/10

70      Solar-Powered Plane Flies for 26 Hours. Alan Cowell. July 8, 2010. *The New York Times.* http://www.nytimes.com/2010/07/09/world/europe/09plane.html retrieved 11/13/10

71      07.07.2010 Solar Impulse HB-SIA has taken off for its first night flight! *Solar Impulse.* http://www.solarimpulse.com/common/documents/news_affich.php?lang=en&group=news&IdArticle=71 retrieved 11/13/10

72      *Solata S380. D.Light.* http://www.dlightdesign.com/products_Solata_S380_global.php retrieved 11/13/10

73      First Solar Signs Contract with PG&E for 300 MW Photovoltaic Solar Power Project. March 09, 2010. *First Solar.* http://investor.firstsolar.com/phoenix.zhtml?c=201491&p=irol-newsArticle&ID=1400401&highlight retrieved 11/13/10

74      Construction of PSE&G's Trenton Solar Farm Underway. August 3, 2010. *PSEG.* http://www.pseg.com/info/media/newsreleases/2010/2010-08-03b.jsp retrieved 11/13/10

75      Baking in the Mojave Sun: US Army Awards $2B Fort Irwin Solar Farm Project. October 15, 2009. *Defense Industry Daily.* http://www.defenseindustrydaily.com/Baking-in-the-Mojave-Sun-US-Army-Awards-2B-Fort-Irwin-Solar-Farm-Project-05858/ retrieved 11/13/10

76	The New Dawn of Solar. *Popular Science*. http://www.popsci.com/popsci/flat/bown/2007/green/item_59.html retrieved 11/13/10

77	New solar panel that magnifies sunlight. *Ecofriend*. http://www.ecofriend.org/entry/new-solar-panel-that-magnifies-the-sunlight/ retrieved 11/13/10

78	Organic activists protest free San Francisco compost. Evelyn Nieves. March 10, 2010. *The Christian Science Monitor*. http://www.csmonitor.com/The-Culture/Gardening/2010/0310/Organic-activists-protest-free-San-Francisco-compost retrieved 11/14/10

79	The 10 riskiest foods in America. *MSNBC*. http://www.msnbc.msn.com/id/33183857/ns/health-food_safety retrieved 11/14/10

80	Eat Greek for Healthier Skin. David A Gabel. August 23, 2010. *Environmental News Network*. http://www.enn.com/health/article/41689 retrieved 11/14/10.

81	Swimmers at sub-tropical beaches show increased risk of illness, study suggests. University of Miami Rosenstiel School of Marine & Atmospheric Science. July 29, 2010. *ScienceDaily*. http://www.sciencedaily.com/releases/2010/07/100728111719.htm retrieved 11/4/10

82	FDA says studies on triclosan, used in sanitizers and soaps, raise concerns. Lindsey Layton. April 8, 2010.*The Washington Post*. http://www.washingtonpost.com/wp-dyn/content/article/2010/04/07/AR2010040704621.html retrieved 11/14/10

83	Learning helps keep brain healthy, UCI researchers find. March 2, 2010. *U.C. Irvine Today*. http://www.today.uci.edu/news/2010/03/nr_gall_100302.php retrieved 11/14/10

84	Key Facts About Seasonal Flu Vaccine. *Centers for Disease Control and Protection*. http://www.cdc.gov/flu/protect/keyfacts.htm retrieved 11/14/10

85	Université Laval. Too much glucosamine can cause the death of pancreatic cells, increase diabetes risk, researchers find. October 27, 2010. *ScienceDaily*. http://www.sciencedaily.com/releases/2010/10/101027111349.htm retrieved 11/14/10

86	10 Weird Uses for Vodka. Brian Clark Howard. *The Daily Green*. http://www.thedailygreen.com/green-homes/latest/vodka-uses-460424 retrieved 11/14/10

87    Walking in the Park Contributes to Happiness. Sue Cartledge. May 12, 2010. *Suite 101*. http://www.suite101.com/content/walking-in-the-park-contributes-to-happiness-a236363 retrieved 11/14/10

88    Ending World Hunger for the Price of a Burger. Stephen B. Tanda. October 18, 2010. *The Huffington Post*. http://www.huffingtonpost.com/stephan-b-tanda/ending-world-hunger-for-t_b_767098.html retrieved 11/21/10

89    PAWS Chicago animal rescue. July, 2009. *Dog Time*. http://dogtime.com/paws-chicago-spotlight.html retrieved 11/15/10

90    The Happy Act: Why Fix Health Care When You Can Subsidize Dog Treats? Howard Gleckman. December 2, 2009. *The Huffington Post*. http://www.huffingtonpost.com/howard-gleckman/the-happy-act-why-fix-hea_b_377534.html retrieved 11/15/10

91    Project of the Month: A Highway Through Key Largo. Debra Wood. Feb 27, 2009. *Engineering News-Record*. http://enr.construction.com/infrastructure/transportation/2009/0227-KeyLargoHighway.asp retrieved 11/15/10

92    Condoms for Endangered Species. Center for Biological Diversity Offers Free Condoms. Feb 11, 2010. *The Huffington Post*. http://www.huffingtonpost.com/2010/02/11/condoms-for-endangered-sp_n_459251.html retrieved 11/15/10

93    Santa Cruz County Health Services Agency Public Health Department-- Homeless Persons Health Project --. http://www.santacruzhealth.org/phealth/2homeless.htm retrieved 11/21/10

94    Food Security: The Challenge of Feeding 9 Billion People. February 12, 2010. *Science Magazine*. http://www.sciencemag.org/cgi/content/full/sci;327/5967/812?maxtoshow=&hits=10&RESULTFORMAT=&fulltext=feed+nine+billion+people+&searchid=1&FIRSTINDEX=0&resourcetype=HWCIT retrieved 11/15/10

95    International Workers Day. Michael Gene Sullivan. April 28, 2010. *The Huffington Post*. http://www.huffingtonpost.com/michael-gene-sullivan/international-workers-day_b_551427.html retrieved 11/15/10

96    History and Origin. *The Holiday Spot*. http://www.theholidayspot.com/mayday/history.htm retrieved 11/15/10

97    Who's the Greenest Generation? New Study Finds Out Who Isn't... Matthew Wheeland. August 28, 2009. *Green Biz*. http://www.greenbiz.com/blog/2009/08/28/whos-greenest-generation-new-study-finds-out-who-isnt#ixzz15NpJIxyQ retrieved 11/15/10

98      How Bisphenol A Products Affect Your Health. Robert Rister. April 13, 2010. *Steady Health.* http://www.steadyhealth.com/articles/How_Bisphenol_A_Products_Affect_Your_Health_a1275.html retrieved 11/21/10

99      Undercover San Francisco Bay Area fish testing finds high mercury levels in supermarket swordfish and tuna. Buffy Martin Tarbox. March 4, 2010. *Got Mercury?* http://www.gotmercury.org/article.php?id=1534 retrieved 11/15/10

100     New Analysis: 15% Cut in U.S. Carbon Emissions Achievable Through Simple Inexpensive Personal Actions. March 12, 2010. *Natural Resources Defense Council.* http://www.nrdc.org/media/2010/100312.asp retrieved 11/15/10

101     Which global consumers are greenest? Not Americans. June 3, 2010. *USA Today.* http://content.usatoday.com/communities/greenhouse/post/2010/06/us-eco-friendly-consumers-17-countries/1 retrieved 11/15/10

102     Climate change affecting Kenya's coffee output. Helen Nyambura-Mwaura. February 11, 2010. *Environmental News Network.* http://www.enn.com/agriculture/article/41006 retrieved 11/15/10

103     Natural Products Association Standard and Certification for Natural Home Care Products. *Natural Products Association.* http://www.npainfo.org/clientuploads/qualityAssurance/Natural_Home_Care/NHC_Standard_020910v01.pdf retrieved 11/15/10

104     Here Come the Electric Cars: "Leaf" and "Volt". Karina Grudnikov. July 31, 2010. *Environmental News Network.* http://www.enn.com/pollution/article/41604 retrieved 11/15/10

105     THINK to begin selling city electric car in New York. April 1, 2010. *Th!nk.* http://www.think.no/nor/Presse/Pressemeldinger/THINK-to-begin-selling-city-electric-car-in-New-York retrieved 11/15/10

106     Walmart Sustainability Index Means Big Business. Tilde Herrera. September 24, 2009. Climate Biz. http://www.greenbiz.com/blog/2009/09/24/walmart-sustainability-index-means-big-business retrieved 11/15/10

107     Coffee Sleeves. *Coffee Shift.* http://coffeeshift.com/article.aspx?id=26 retrieved 11/15/10

108     Reusable Coffee Sleeves. January 8, 2010. *Daily Danny.* http://www.dailydanny.com/?p=2511 retrieved 11/15/10

109     U.S. farmers can't meet booming corn demand. Charles Abbott. July 10, 2010. *Environmental News Network*. http://www.enn.com/agriculture/article/41526 retrieved 11/15/10

110     Biofuels 'done right' can curb greenhouse gas emissions and provide other benefits. Chris Emery. July 16, 2009. *Princeton University*. http://www.princeton.edu/main/news/archive/S24/77/72I91/index.xml?section=topstories retrieved 11/15/10

111     Kew's Millennium Seed Bank partnership 'top banana' as it celebrates banking 10% of the world's wild plant species. October 15, 2009. *Kew*. http://www.kew.org/news/kew-millennium-seed-bank-partnership-top-banana-celebrate-banking-10-percent.htm retrieved 11/15/10

112     The swelling McMansion backlash. Christopher Solomon. *MSN Real Estate*. http://realestate.msn.com/article.aspx?cp-documentid=13107733 Retrieved 11/17/10

113     Switchgrass Yields Energy-efficient, Clean Fuel, Study Finds. January 11, 2008. *PBS News Hour*. http://www.pbs.org/newshour/updates/science/jan-june08/switchgrass_01-11.html retrieved 11/17/10

114     University of California - Irvine (January 22, 2010). Urban 'green' spaces may contribute to global warming. *ScienceDaily*. http://www.sciencedaily.com/releases/2010/01/100119133515.htm retrieved 11/17/10

115     Green With Guilt. George F. Will. June 4, 2009. *The Washington Post*. http://www.washingtonpost.com/wp-dyn/content/article/2009/06/03/AR2009060303240.html retrieved 11/17/10

116     Difference Between Earth Day & Arbor Day. Lexa W. Lee. December 22, 2009. *eHow*. http://www.ehow.co.uk/facts_5782429_difference-earth-day-arbor-day.html retrieved 11/17/10

117     Charts and Tables from Clean Tech Trends 2010. *Clean Edge*. http://www.cleanedge.com/reports/charts-reports-jobtrends2010.php retrieved 11/17/10

118     Wasted Light means Wasted Energy. *Light Pollution*. http://www.lightpollution.org.uk/index.php?pageId=5 retrieved 11/17/10

119     Want some kinetic energy with those fries? Sharon Vaknin. July 8, 2009. *CNet News*. http://news.cnet.com/8301-11128_3-10282501-54.html retrieved 11/17/10

120    Tapping the Earth for home heating and cooling. Martin LaMonica. January 14, 2009. *CNet News.* http://news.cnet.com/8301-11128_3-10131539-54.html#ixzz15wyg3z4m retrieved 11/21/10

121    Earth Hour 2009: Lights Out In 84 Countries. Caryn Rousseau. March 27, 2009. *The Huffington Post.* http://www.huffingtonpost.com/2009/03/27/earth-hour-2009-lights-ou_n_179895.html retrieved 11/17/10

122    The Secret to Turning Consumers Green. Stephanie Simon. October 18, 2010. *The Wall Street Journal.* http://online.wsj.com/article/SB10001424052748704575304575296243891721972.html?mod=WSJ_business_LeftSecondHighlights retrieved 11/17/10

123    The Bloom Box: An Energy Breakthrough? February 18, 2010. *CBS News.* http://www.cbsnews.com/stories/2010/02/18/60minutes/main6221135.shtml retrieved 11/17/10

124    Navajos Hope to Shift From Coal to Wind and Sun. Mireya Navarro. October 25, 2010. *The New York Times.* http://www.nytimes.com/2010/10/26/science/earth/26navajo.html?pagewanted=1&_r=1 retrieved 11/18/10

125    LIFE: Clean Energy from Nuclear Waste. *National Ignition Facility.* https://lasers.llnl.gov/about/missions/energy_for_the_future/life/ retrieved 11/18/10

126    Norway opens world's first osmotic power plant. November 24, 2009. *CNet News.* http://news.cnet.com/8301-11128_3-10404158-54.html retrieved 11/18/10

127    Good data needed for small-wind turbines to spin. Martin LaMonica. May 12, 2010. *CNet News.* http://news.cnet.com/8301-11128_3-20004716-54.html#ixzz15c6HXyrn retrieved 11/18/10

128    REN21: More than half of new power in U.S., EU is green. Reuters. July 16, 2010. *CNet News* http://news.cnet.com/8301-11128_3-20010770-54.html#ixzz15c7Ajj9g retrieved 11/18/10

129    34 miles per gallon: The new normal. Peter Valdes-Dapena. April 1, 2010. *CNN Money.* http://money.cnn.com/2010/04/01/autos/cafe_standards_final/index.htm retrieved 11/18/10

130    A Step Closer to Safer Food. Rep. Louise Slaughter. July 30, 2009. *The Huffington Post.* http://www.huffingtonpost.com/rep-louise-slaughter/a-step-closer-to-safer-fo_b_248258.html retrieved 11/18/10

131    Bill would restrict antibiotics in food animals. Stacy Finz. July 31, 2009. *SF Gate*. http://articles.sfgate.com/2009-07-31/news/17219549_1_antibiotics-animals-meat-and-poultry retrieved 11/18/10

132    Climate Change Indicators in the United States. *U.S. Environmental Protection Agency*. http://www.epa.gov/climatechange/indicators.html retrieved 11/18/10

133    Cleansing Soap. September 2010. *National Conference of State Legislatures*. http://www.ncsl.org/?tabid=21077 retrieved 11/18/10

134    Kerry and Lieberman Release Draft Climate Bill While EPA Issues Final Tailoring Rule. Andrea Carruthers, et al. May 21, 2010. *Environmental Leader*. http://www.environmentalleader.com/2010/05/21/kerry-and-lieberman-release-draft-climate-bill-while-epa-issues-final-tailoring-rule/ retrieved 11/18/10

135    Obama's $8 billion plan would dramatically shorten trips from Chicago to other Midwest cities. Jon Hilkevitch. April 17, 2009. *The Chicago Tribune*. http://www.chicagotribune.com/news/local/chi-high-speed-rail-17-apr17,0,1876557.story retrieved 11/18/10

136    Supreme Court Eases Campaign Finance Curbs. Nina Totenberg. January 21, 2010. *NPR*. http://www.npr.org/templates/story/story.php?storyId=122823090 retrieved 11/18/10

137    Clean Air Act. *U.S. Environmental Protection Agency*. http://www.epa.gov/air/caa/ retrieved 11/18/10

138    Obama Creates National Ocean Council to Oversee Protection of Our Oceans, Coasts & Great Lakes. Matthew McDermott. July 20, 2010. *Treehugger*. http://www.treehugger.com/files/2010/07/obama-creates-national-ocean-council-executive-order.php retrieved 11/18/10

139    Exxon Sinks $600M Into Algae-Based Biofuels in Major Strategy Shift. Katie Howell. July 14, 2009. *The New York Times*. http://www.nytimes.com/gwire/2009/07/14/14greenwire-exxon-sinks-600m-into-algae-based-biofuels-in-33562.html retrieved 11/18/10

140    Survey: Consumers intrigued by electric cars. Martin LaMonica. March 19, 2010. *CNet News*. http://news.cnet.com/8301-11128_3-20000772-54.html#ixzz15eZXDeGw retrieved 11/18/10

141     Paving the way for greener asphalt. Elsa Wenzel. May 29, 2008. *CNet News.* http://news.cnet.com/8301-11128_3-9955323-54.html?tag=mncol;1n retrieved 11/21/10

142     Boeing and Industry Study Shows Biofuels Perform Effectively as Jet Fuel. *BOEING.* http://boeing.mediaroom.com/index.php?s=43&item=714 retrieved 11/18/10

143     Halving Emissions by 2050 - Aviation Brings its Targets to Copenhagen. December 8, 2009. *International Air Transport Association.* http://www.iata.org/pressroom/pr/Pages/2009-12-08-01.aspx retrieved 11/18/10

144     Ford Transit Connect Electric: Impressively unimpressive. Antuan Goodwin. March 9, 2010. *CNet News.* http://reviews.cnet.com/8301-13746_7-10466379-48.html?tag=mncol;8n retrieved 11/18/10

145     Quebec utility bets on the electric car. Bertrand Marotte. August 29, 2010. *The Globe and Mail.* http://www.theglobeandmail.com/report-on-business/quebec-utility-bets-on-the-electric-car/article1689394/ retrieved 11/18/10

146     Earth Day Special: The Human Car HC Imagine_PS. Aaron Turpen. *Future Cars.* http://www.futurecars.com/future-cars/electric-cars/earth-day-special-the-human-car-hc-imagineps retrieved 11/18/10

147     Study: Fleet Buyers Can Jump-Start Plug-in Vehicles. Martin LaMonica. November 15, 2010. *CNet News.* http://news.cnet.com/8301-11128_3-20022796-54.html retrieved 11/18/10

148     [147] Body Heat: Sweden's New Green Energy Source. Tara Kelly. April 15, 2010. *Time.* http://www.time.com/time/health/article/0,8599,1981919,00.html retrieved 11/18/10

149     Honda to Lease Electric Scooters From December. Yuri Kageyama. April 13, 2010. *ABC News/Money.* http://abcnews.go.com/Business/wireStory?id=10358147 retrieved 11/18/10

150     Case Study: Chicago City Hall's Green Roof. Kelly. April 21, 2010. *The Metropolitan Field Guide.* http://www.metrofieldguide.com/?p=82 retrieved 11/18/10

151     The Benefits of Sedum Green Roofs – Have a living roof. Louise Hampson. May 12, 2010. *Suite101.* http://www.suite101.com/content/the-benefits-of-sedum-green-roofs-a237510#ixzz15gxxl3Iz retrieved 11/18/10

152     Green Tires From Orange Peels Use Less Oil, Bring Better MPGs. John Voelcker. December 10, 2009. *Green Car Reports*. http://www.greencarreports.com/blog/1039964_green-tires-from-orange-peels-use-less-oil-bring-better-mpgs retrieved 11/18/10

153     The Benefits of Composting Toilets. *Envirolet composting toilet world*. http://compostingtoilet.org/compost_toilets_explained/the_benefits_of_composting_toilets/index.php retrieved 11/18/10

154     Synthetic 'tree' promises to catch carbon. Sharon Vaknin. June 24, 2009. *CNet News*. http://news.cnet.com/8301-11128_3-10272020-54.html retrieved 11/18/10

155     Trailer-mounted wind turbine tested at Army base. Martin LaMonica. April 8, 2010. *CNet News*. http://news.cnet.com/8301-11128_3-20002014-54.html?tag=mncol;1n retrieved 11/18/10

156     Don't go off-road in electric car made of bamboo. Tim Hornyak. May 21, 2010. *CNet News*. http://news.cnet.com/8301-17938_105-20005629-1.html#ixzz15hXC0k1x retrieved 11/18/10

157     Amtrak Biodiesel Heartland Flyer named on the year's 50 biggest and coolest breakthroughs. Heartland Flyer. http://www.heartlandflyer.com/news-special-events/detail.aspx?id=73 retrieved 11/18/10

158     Cheetos bags find new life as MP3 speakers. Tim Hornyak. October 24, 2009. *CNet News*. http://news.cnet.com/8301-17938_105-10382623-1.html#ixzz15hcjPVZS retrieved 11/18/10

159     Car is flushed with power. *Geneco Sustainable Solutions*. http://www.geneco.uk.com/about/index.aspx?id=6028 retrieved 11/18/10

160     Turn your office expense reports into toilet paper. Tim Hornyak. February 6, 2010. *CNet News*. http://news.cnet.com/8301-17938_105-10382622-1.html#ixzz15heLWVhc retrieved 11/18/10

161     The Vertical Farm. *The Vertical Farm Project*. http://www.verticalfarm.com/more retrieved 11/18/10

162     Old printer cartridges turned into bike path. Justin Yu. June 14, 2010. *CNet News*. http://news.cnet.com/8301-17938_105-20007655-1.html retrieved 11/18/10

163     SkySails - turn wind into profit. *Sky Sails*. http://www.skysails.info/english/ retrieved 11/18/10

164     Orange's Prototype Power-Welly Boots Are Made For Charging. Rebecca Boyle. June 7, 2010. *Popsci*. http://www.popsci.com/gadgets/article/2010-06/these-boots-are-made-charging retrieved 11/19/10

165     Congratulations to Virginia Tech and Solar Decathlon Europe. June 27, 2010. *U.S. Department of Energy.* http://www.solardecathlon.gov/blog/archives/tag/solar-decathlon-europe-2010 retrieved 11/19/10